T0135753

# Linear and Nonlinear Model Order Reduction for Numerical Simulation of Electric Circuits

**Kasra Mohaghegh**

Bergische Universität Wuppertal
Fachbereich Mathematik und Naturwissenschaften
Lehrstuhl für Angewandte Mathematik
und Numerische Mathematik

Ph.D. thesis

Bibliographic information published by the Deutsche Nationalbibliothek

The Deutsche Nationalbibliothek lists this publication in the Deutsche Nationalbibliografie; detailed bibliographic data are available in the Internet at http://dnb.d-nb.de .



ISBN 978-3-8325-2711-2

Logos Verlag Berlin GmbH
Comeniushof, Gubener Str. 47,
D-10243 Berlin
Tel.: +49 (0)30 42 85 10 90
Fax: +49 (0)30 42 85 10 92
INTERNET: http://www.logos-verlag.de

# Acknowledgment

First, I thank my both supervisors Prof. Dr. Roland Pulch and Dr. Jan ter Maten, for their continuous support in my Ph.D. program. Roland was always there to listen and to give advice. He taught me how to ask questions and express my ideas. He patiently corrected all my academic writings. He showed me different ways to approach a research problem and to be precise. It was Jan from whom I learned how a scientific research should be organized. He taught me during my stay at NXP the basic stuff in model order reduction. Also in the first place I thank Prof. Dr. Michael Günther the coordinator of the coupled multiscale simulation and optimization in nanoelectronics (COMSON) project[1] who was responsible for involving me in the COMSON project. He was always available for listening the problems and helped me in solving them. Michael is always ready for scientific discussion and kindly commented on all of my presentations.

A special thank goes to my co-supervisor, Dr. Michael Striebel, who is most responsible for helping me complete the writing of this dissertation as well as the challenging research that lies behind it. Michael has been a friend and mentor. He taught me how to write academic papers, made me a better programmer, had confidence in me when I doubted myself, and brought out the good ideas in me. Without his encouragement and constant guidance I could not have finished this dissertation. He was always there to meet and talk about my ideas, to proofread and mark up my papers and chapters, and to ask me good questions to help me think through my problems (whether philosophical, analytical or computational). As I cannot understand "deutsch" he always writes and telephones for me, although regarding this topic I owed to almost every colleague in this group that patiently help me regarding my lack of knowledge in German language.

Besides my supervisors, I thank the rest of my colleagues and friends in the COMSON project due to quite long list I am not able to mention all of them here. During the past three years we met in several meetings and had wonderful discussions. Specially I want to thank Zoran Ilieveski from TU/e Eindhoven, The Netherlands, for all nice discussions which we had regarding MOR. Let me also say "thank you" to the following colleagues at Bergische Universität Wuppertal which make Wuppertal a nice place to

[1] COMSON is a Marie Curie research training network supported by the European commission in the framework of the program "structuring the European research area" within the 6th framework research program of the European Union.

work: Dr. Andreas Bartel, Dr. Massimiliano Culpo, Patrick Deuß, Sebastian Schöps, Dr. Renate Winkler, Dr. Markus Brunk, Kirill Gordine (system administrator) and Elvira Mertens (secretary of the group). I also should thank Yao Yue and Prof. Dr. Karl Meerbergen from Katholieke Universiteit Leuven, Belgium, for sharing there codes and papers for parametric model order reduction PIMTAB.

Last, but not the least, I thank my family: my parents, Ali A. Mohaghegh, and Zahra Shahmirza, for giving me life in the first place, for educating me with aspects from both arts and sciences, for unconditional support and encouragement to pursue my interests, even when the interests went beyond boundaries of language, field and geography. My brother Kambiz do, for permanent support and kindness, for reminding me that he is always available to help me. My sister in law Leila Mottaghi do, for her patience and support. My sister Parisa and her husband, Mehrdad, do, for listening to my complaints and frustrations and for believing in me and finally to my lovely wife Maryam Saadvandi do, for everything that only she and I know, for endless support and love, for reminding me what really matters and for always being there to 100 percent!

تقدیم به پدرم به خاطر یک عمر تلاشش

تقدیم به مادرم به خاطر تمام رنجهایش

تقدیم به مریم به خاطر زلالیش

تقدیم به برادرم به خاطر قلب مهربانش

تقدیم به خواهرم به خاطر حضور خالصانه اش

تقدیم به لیلا به خاطر صبرش

و باز هم تقدیم به مریم

به خاطر اینکه هیچ چیز با ارزش تر از یک دوستی واقعی نیست.

# Contents

**Motivation** **1**

**1 Introduction** **5**

1.1 Linear Dynamical Systems . . . . . . . . . . . . . . . . . . . . . . . . . . 5

1.1.1 External representation . . . . . . . . . . . . . . . . . . . . . . 6

1.1.2 Internal representation . . . . . . . . . . . . . . . . . . . . . . . 7

1.1.3 Reachability . . . . . . . . . . . . . . . . . . . . . . . . . . . . . 10

1.1.4 Observability . . . . . . . . . . . . . . . . . . . . . . . . . . . . 12

1.1.5 Infinite gramians . . . . . . . . . . . . . . . . . . . . . . . . . . 13

1.1.6 Realization problem . . . . . . . . . . . . . . . . . . . . . . . . 14

1.1.7 Hankel operator . . . . . . . . . . . . . . . . . . . . . . . . . . . 15

1.2 Differential Algebraic Equations . . . . . . . . . . . . . . . . . . . . . . 16

**2 Linear model order reduction** **25**

2.1 Linear reduction of dynamical systems . . . . . . . . . . . . . . . . . . 25

2.2 MOR Methods . . . . . . . . . . . . . . . . . . . . . . . . . . . . . . . . 27

2.2.1 Introduction to Krylov Projection Techniques . . . . . . . . . . 28

2.2.2 Methods based on Hankel norm approximations and TBR . . . 32

2.3 Homotopized DAEs . . . . . . . . . . . . . . . . . . . . . . . . . . . . . 35

2.4 Model Order Reduction and $\varepsilon$-Embedding . . . . . . . . . . . . . . . . 36

2.5 General Linear Systems . . . . . . . . . . . . . . . . . . . . . . . . . . . 41

2.5.1 Transformation to Kronecker Form . . . . . . . . . . . . . . . . 41

2.5.2 Transformation via Singular Value Decomposition . . . . . . . . 44

2.6 Parametric Model Order Reduction . . . . . . . . . . . . . . . . . . . . 46

**3 Nonlinear model order reduction** **51**
3.1 Nonlinear versus Linear MOR . . . 51
3.2 POD and Adaptations . . . 53
3.2.1 Basis of POD . . . 53
3.2.2 Adaption of POD for nonlinear problems . . . 54
3.3 Trajectory Piecewise Linear Techniques . . . 58
3.3.1 Selection of linearization points . . . 58
3.3.2 Determination of weights . . . 59
3.3.3 Reduction of linear sub-models . . . 60
3.3.4 Construction of reduced order basis . . . 61
3.4 Balanced Truncation in Nonlinear MOR . . . 61

**4 Examples and Numerical Results** **63**
4.1 Linear Circuits . . . 63
4.2 Nonlinear Circuits . . . 76

**5 Conclusions and outlook** **87**

**Bibliography** **89**

*When you look into the abyss,*
*the abyss also looks into you.*
Friedrich Nietzsche

# Motivation

Integrated circuits (ICs) are used in almost all electronic equipment in use today and have revolutionized the world of electronics. Nowadays ICs comprise about hundred millions of transistors on slightly more than one square centimeter and some ten thousand ICs (from one single transistor to complex circuits) are produced on silicon wafers with a diameter of 45nm. The ever decreasing feature size of ICs has lead to structures on the nanoscale. Current CPUs are produced using 45nm technology. We are on the verge of 32nm technology and 22nm technology is under development. Although we have not fully reached nanoelectronics, which is often defined to start at 11nm, we are facing nanoscale structures.

This development is accompanied by increasing numbers of devices and circuit elements as well as packaging density. What Gordon E. Moore[1] predicted and what is known under the name of *Moore's Law*[2] is still valid: the number of transistors in one IC doubles in every 18-24 months and thus doubles the speed of computations, see Figure 1. The tendency to analyze and construct systems of ever increasing complexity is becoming more and more a dominating factor in progress of chip design. Along with this tendency, as we already mentioned, the complexity of the mathematical models increases both in structure and dimension. Complex models are more difficult to analyze, and due to this it is also harder to develop control algorithms. Therefore model order reduction (MOR) is of utmost importance.

For linear systems, quite a number of MOR approaches are well-established and have proved to be very useful in the case of ordinary differential equations (ODEs). On the other hand, we deal with differential-algebraic equations (DAEs), which result from models of electronic circuits based on network approaches. There are the direct and the indirect strategy to convert a semi-explicit DAE of index 1 into an ODE. We apply the direct approach, where an artificial parameter is introduced in a linear system of DAEs. It follows a singularly perturbed problem. This artificial parameter is handled as a system parameter and so the parametric MOR is used. Numerical simulations for linear test cases are presented. The first two examples are studied with general MOR

[1] http://en.wikipedia.org/wiki/Gordon_E._Moore.

[2] Moore's law describes a long-term trend in the history of computing hardware, in which the number of transistors that can be placed inexpensively on an integrated circuits has doubled approximately every two years.

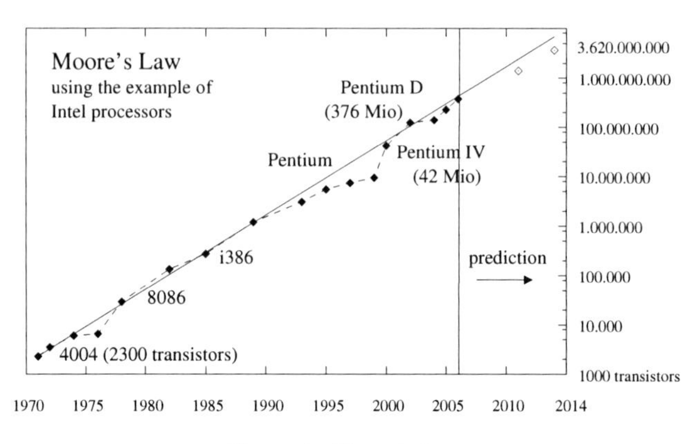


Figure 1: Moore's Law.

methods while the last two examples are regarding the direct approach.

Meanwhile accurate models for MOS-devices introduce highly nonlinear equations. And, as the packing density in circuit design is growing, very large nonlinear systems arise. Hence, there is a growing request for reduced order modeling of nonlinear problems. In this thesis we review the status of existing techniques for nonlinear model order reduction by investigating how well these techniques perform for circuit simulation. In particular the Trajectory PieceWise Linear (TPWL) method and the Proper Orthogonal Decomposition (POD) approach are taken under consideration. The reduction of a nonlinear transmission line with TPWL and POD methods and the reduction of an inverter chain with TPWL are presented.

This work is organized as follows:

Ch.1- The control theory plays a major role in the analysis of model order reduction. We will shortly study some fundamental definitions and aspects of it. As the electrical circuits are described by DAEs, their basic properties are also discussed.

Ch.2- This part is solely dedicated to linear methods. We start with discussing two classical linear methods and then mention some new techniques. The direct approach is used to handle the DAEs and suitable reduction possibilities are studied. The methods first discussed for semi-explicit systems are then extended to a general linear case. Some popular parametric reduction schemes are reviewed at the end of this part, where the PIMTAB method is discussed in more details.

Ch.3- This part is dedicated to nonlinear MOR strategies. The problems handling the nonlinearities in reductions are briefly discussed. Some practical methods such as POD and TPWL approach are studied in details. To have a general overview of almost all available techniques also some other techniques which are not applicable for circuit simulation are also introduced.

Ch.4- This chapter includes the numerical simulations and the discussion of results. Based on the structure of this thesis we have two main parts for simulation. In the first part we have linear test examples and in the second part we consider nonlinear examples. In the linear part we study the circuits in frequency domain whereas in nonlinear parts we analyze the examples in time domain.

# Chapter 1

# Introduction

In this chapter we briefly discuss the main basic concept of the control theory, differential-algebraic equations (DAEs) and finally close this part with index definitions. This introduction surveys basic concepts and questions of the theory and describes some typical examples. The first part is devoted to structural properties of the linear dynamical systems. It contains basic results in reachability, observability, realization and the Hankel operator. The modified nodal analysis (MNA) and the basic types of DAEs are introduced in the second part. Finally four different index concepts will be defined and explained with examples. General references for the material in this introduction are [1, 56, 60, 76].

## 1.1 Linear Dynamical Systems

Metaphorically speaking, a dynamical system describes, by means of a so called *evolution function* how a point's position changes with an evolution parameter to which one usually assigns time. A dynamical system is called linear when its evolution function is linear. We are interested in dynamical systems that map an input $u$ to an output $y$.

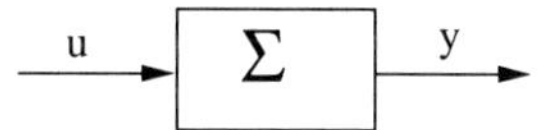


Figure 1.1: The input-output behavior of a dynamical system.

We call the system under consideration $\Sigma$ and distinguish two different representations of $\Sigma$.

- External representation,
- Internal representation.

In the *external representation* the output is given as a function of the input. Regarding physical systems, the inner structure of the object under consideration is hidden, it is given as a black box only. The *internal representation* on the other hand introduces the state $x$ as additional quantity. Here, going from input to output is a two step process in general. Nifow we discuss some properties of both descriptions and we shall also address the equivalence of the representations. We are mainly interested in continuous time problems where the evolution parameter $t$ is chosen from the set of real numbers. However, to illustrate certain issues we use discrete time formulation, where $t$ is also an integer.

### 1.1.1 External representation

A linear system can be viewed as a linear operator $S$ that maps from the input space into the output space:

$$S : \{u : D_t \to \mathbb{R}^m\} \to \{y : D_t \to \mathbb{R}^p\},$$
$$u \mapsto y = S(u).$$

$S$ can be expressed by an integral for continuous time problems and by a sum for discrete time problems.

1. Continuous time problem:

$$y(t) = \int_{-\infty}^{\infty} h(t, \tau)u(\tau)d\tau, \quad t \in D_t, \quad D_t = \mathbb{R}. \tag{1.1}$$

2. Discrete time problem:

$$y(t) = \sum_{\tau \in \mathbb{Z}} h(t, \tau)u(\tau), \quad t \in D_t, \quad D_t = \mathbb{Z}. \tag{1.2}$$

Where a matrix valued function is defined by: $h : D_t \times D_t \to \mathbb{R}^{p\times m}$ and $h(t, \tau)$ is called the kernel or waiting pattern of $\Sigma$. The system $\Sigma$ is called *causal* if $h(t, \tau) = 0$ for $t \leq \tau$, i.e. future input does not effect present output and called *time*[*-shift*]*-invariant* if $h(t, \tau) = h(t - \tau)$.

Therefore, for linear time-invariant (LTI) systems $S$ is a *convolution operator*

$$S : u \to y = S(u) = h * u,$$

with (continuous time case)

$$(h * u)(t) = \int_{-\infty}^{\infty} h(t - \tau)u(\tau)d\tau.$$

It is assumed from now on that $S$ is both causal and time-invariant. In addition we will express the output as a sum of two terms, the instantaneous and the dynamic action, as follows:

$$y(t) = \underbrace{h_0 u(t)}_{\text{instantaneous action}} + \underbrace{\int_{-\infty}^{\infty} h_a(t-\tau)u(\tau)d\tau,}_{\text{dynamic action}}$$

where $h_0 \in \mathbb{R}^{p\times m}$ and $h_a$ is a smooth kernel. Then $h$, the *impulse response* of $\Sigma$, can be expressed as

$$h(t) = h_0\delta(t) + h_a(t), \quad t \geq 0, \tag{1.3}$$

with $\delta$ being the Dirac delta function and $h_a$ is an analytic function. Then, $h_a$ is uniquely determined by the coefficients of its Taylor series expansion at $t = 0$:

$$h_a(t) = h_1 + h_2\frac{t}{1!} + h_3\frac{t^2}{2!} + \cdots + h_k\frac{t^{k-1}}{(k-1)!} + \cdots, \quad h_k \in \mathbb{R}^{p\times m}.$$

It follows that if (1.3) is satisfied, the output $y$ is at least as smooth as the input $u$, and $\Sigma$ is consequently, called a *smooth* system. Hence smooth continuous-time linear systems can be described by means of the infinite sequence of the $p \times m$ matrices $h_i, i \geq 0$. We formalize this conclusion next.

**Definition 1.1.** *The external description of a causal LTI-system with $m$ inputs and $p$ outputs is given by an infinite series of $p \times m$ matrices*

$$(h_0, h_1, h_2, \ldots, h_k, \ldots), \quad h_k \in \mathbb{R}^{p\times m}.$$

*The matrices $h_k$ are often referred to as the Markov parameters of the system.*

Frequently, an LTI system is characterized in the frequency domain, instead. The transition from time to frequency domain is given by the Laplace transformation (Z-transform for discrete time systems). The Laplace transform of the impulse response $h(t)$ is called *transfer function* and can be given as:

$$H(s) = h_0 + h_1 s^{-1} + h_2 s^{-2} + \cdots + h_k s^{-k} + \cdots.$$

Figure 1.2 shows the relation from time-domain to frequency-domain. Therefore, (1.1) and (1.2) can be written as

$$Y(s) = H(s)U(s),$$

for more details see [51]. We will discuss this procedure in the next chapter in detail.

### 1.1.2 Internal representation

Alternatively, we can characterize a linear system via its *internal description*, see [51], which in addition to the input $u$ and the output $y$ uses the state $x$. In the internal description, the *state*

$$x : D_t \to \mathbb{R}^n,$$

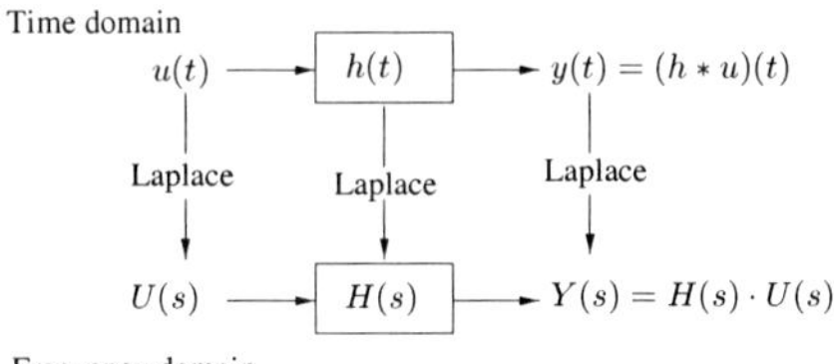


Figure 1.2: The input-output behavior of a dynamical system in time domain and frequency domain.

is introduced, and the mapping from input $u$ to output $y$ becomes a two stage process. The first stage process is the *state equation* which could be introduced both in continuous- and discrete-time domain as follows:

- Continuous time:

$$\dot{x}(t) = Ax(t) + Bu(t), \quad t \in D_t,\ D_t = \mathbb{R}, \tag{1.4a}$$

- Discrete time:

$$x(t+1) = Ax(t) + Bu(t), \quad t \in D_t, D_t = \mathbb{Z}, \tag{1.4b}$$

with $A \in \mathbb{R}^{n \times n}$ and $B \in \mathbb{R}^{n \times m}$. The second stage process is the *output equation*:

$$y(t) = Cx(t) + Du(t)$$

with $C \in \mathbb{R}^{p \times n}, \quad D \in \mathbb{R}^{p \times m}$. A linear system in *internal* or *state space* description is the quadruple of linear maps

$$\Sigma = \left( \begin{array}{c|c} A & B \\ \hline C & D \end{array} \right).$$

The *dimension* of the system is defined as the dimension of the associated state space $\dim \Sigma = n$.

**Definition 1.2.** *$\Sigma$ is called "stable" if for all $\lambda \in \Lambda(A)$:*

| | |
|---|---|
| *continuous time:* | $\mathrm{Re}(\lambda) < 0,$ |
| *discrete time:* | $\lvert\lambda\rvert < 1,$ |

*where $\Lambda(A)$ is the set of all eigenvalues of $A$ and is called the "spectrum" of $A$.*

For a given initial value $x_0 = x(t_0)$ for the state $x$ at time $t_0$ and a given input $u$, the solution of the state-space equations is, for the continuous-time problem:

$$x(t) = \phi(u; x_0; t) = e^{A(t-t_0)}x_0 + \int_{t_0}^{t} e^{A(t-\tau)}Bu(\tau)d\tau.$$

where the matrix exponential is defined by the series:

$$e^{At} = I_n + \frac{t}{1!}A + \frac{t^2}{2!}A + \cdots + \frac{t^k}{k!}A^k + \cdots .$$

Therefore, the output is given by

$$\begin{aligned} y(t) &= C\phi(u; x(t_0); t) + D \cdot u(t) \\ &= Ce^{A(t-t_0)}x(t_0) + \int_{t_0}^{t} Ce^{A(t-\tau)}Bu(\tau)d\tau + Du(t) \\ &= C\phi(0; x(t_0); t) + C\phi(u; 0; t) + Du(t). \end{aligned}$$

The *impulse response* describes the reaction of the system as a function of time (or possibly as a function of some other independent variable that parameterizes the dynamic behavior of the system). For $t_0 \to -\infty$ and $x_0 = 0$ the output becomes:

$$\begin{aligned} y(t) &= Du(t) + \int_{-\infty}^{t} Ce^{A(t-\tau)}Bu(\tau)d\tau \\ &= \int_{-\infty}^{t} h(t-\tau)u(\tau)d\tau \end{aligned}$$

where $h(t)$ is *impulse response.* If we compare the above expressions with (1.1) and (1.2), it follows that the impulse response $h$ has the form:

$$h(t) = \begin{cases} Ce^{At}B + D \cdot \delta(t) & t \geq 0, \\ 0 & t < 0. \end{cases}$$

As we know from the external description, the transfer function of $\Sigma$ is the Laplace transform of $h$

$$H(s) = \mathbb{L}(h(t)) = D + C(sI - A)^{-1}B.$$

Then with basic calculation we have:

$$\begin{aligned} H(s) &= D + \tfrac{1}{s}C(I - \tfrac{1}{s}A)^{-1}B \\ (A \text{ is stable}) &\overset{\text{Neumann Series}}{=} D + \tfrac{1}{s}C(I + \tfrac{1}{s}A + \tfrac{1}{s^2}A^2 + \cdots + \tfrac{1}{s^k}A^k + \cdots)B \\ &= D + CBs^{-1} + CABs^{-2} + \cdots + CA^{k-1}Bs^{-k} + \cdots . \end{aligned}$$

We need the stability of $A$ for the convergence of Neumann series. Hence, the Markov parameters of $\Sigma = \left(\begin{array}{c|c} A & B \\ \hline C & D \end{array}\right)$ are given by:

$$(D, CB, CAB, \ldots, CA^{k-1}B, \ldots)$$

and we have found the external description.

### 1.1.3 Reachability

Now, we want to investigate, to which extend the state $x(t)$ of the system can be manipulated through the input $u$. In this consideration only the state space equation is involved. Therefore, we ignore the output stage and consider the first part of the linear system only.

**Definition 1.3.** *Given* $\Sigma = \left(\begin{array}{c|c} A & B \\ \hline & \end{array}\right)$, $A \in \mathbb{R}^{n\times n}$, $B \in \mathbb{R}^{n\times m}$. *A state* $\bar{x} \in \mathbb{R}^n$ *is "reachable" from the zero state if there exists an input function* $\bar{u}$ *of finite energy* [1] *and a time* $\bar{T} < \infty$ *such that*

$$\bar{x} = \phi(\bar{u}; 0; T).$$

The *reachable subspace* $\mathscr{R} \subseteq \mathbb{R}^n$ of $\Sigma$ is the set containing all reachable states of $\Sigma$. We say that $\Sigma$ is *completely reachable* if $\mathscr{R} = \mathbb{R}^n$ holds.

Consider the discrete-time state equations in (1.4b) with $x(0) = x_0 = 0$ then we have:

- The first step: $x(1) = Bu(0)$ i.e., reachable in one step: $\mathrm{im}(B)$.
- The second step: $x(2) = Ax(1) + Bu(1) = ABu(0) + Bu(1)$ i.e., reachable in two steps: $\mathrm{im}(B, AB)$.

  $\vdots$

- The $k$th step: $x(k) = A^{k-1}Bu(0) + A^{k-2}Bu(1) + \cdots + ABu(k-2) + Bu(k-1)$ i.e., reachable in $k$ steps: $\mathrm{im}(B, AB, \ldots, A^{k-1}B)$.

This motivates, the following result for $k \to \infty$.

**Theorem 1.1.** *Given* $\Sigma = \left(\begin{array}{c|c} A & B \\ \hline & \end{array}\right)$ *then for both continuous-time and discrete-time case,* $\mathscr{R}$ *is a linear subspace of* $\mathbb{R}^n$, *given by the formula*

$$\mathscr{R} = \mathrm{im}R(A, B),$$

*with* $R(A, B) = (B, AB, \ldots, A^{k-1}B, \ldots)$ *is the reachability matrix of* $\Sigma$.

These results follow from Theorem 1.1:

- Due to the Cayley-Hamilton theorem the image of $R$ is determined at most by the first $n$ terms $A^kB$, $k = 0, 1, \ldots, n-1$.
- $\mathrm{im}R(A, B)$ is also known as the Krylov subspace that is generated by $A$ and $B$.

[1] A finite energy signal is a signal $f(t)$ for which $\int_{-\infty}^{\infty} |f(t)|^2\, dt < \infty$.

The concept of the *reachability gramian* is very useful to prove Theorem 1.1.

**Definition 1.4.** *The finite reachability gramian at time $t < \infty$ is defined as:*

$$\underbrace{P = \int_0^t e^{A\tau} BB^* e^{A^*\tau} d\tau}_{\text{continuous time systems}} \quad | \quad \underbrace{P = \mathscr{R}_t(A,B)\mathscr{R}_t^*(A,B)}_{\text{discrete time systems}}. \tag{1.5}$$

The reachability gramian (1.5) has the following remarkable properties:

(i) $P(t) = P^*(t)$ is positive semi-definite,

(ii) $\text{im } P(t) = \text{im } R(A,B)$ for all $t > 0$ (continuous-time systems).

The second statement above says that the system can be steered to any reachable state arbitrarily fast. The proof of Theorem 1.1 constructs the *cheapest* input function $\bar{u}$ that drives $\Sigma$ to the state $\bar{x} \in \mathbb{R}$ in time $\bar{T}$. In the following we give the scratch of the proof of Theorem 1.1:

i) $\mathscr{R}$ is linear subspace: $x_1, x_2 \in \mathscr{R} \Rightarrow \alpha x_1 + \beta x_2 \in \mathscr{R}$.

ii) $\mathscr{R} \subseteq \text{im } R(A,B)$ : use $e^{At} = \sum_{i>0} \frac{t^{i-1}}{(i-1)!} A^{i-1}$.

iii) $\text{im } R(A,B) \subseteq \mathscr{R}$: $\bar{x} \in \text{im } R(A,B) \Rightarrow \bar{x} = P(\bar{T})\ \bar{\xi}$ for any $\bar{T} > 0$, choose

$$\bar{u}(t) = B^* e^{A^*(\bar{T}-t)}\bar{\xi},$$

so we have:

$$\begin{aligned} \phi(\bar{u}; 0; \bar{T}) &= \int_0^{\bar{T}} e^{A(\bar{T}-\tau)} BB^* e^{A^*(\bar{T}-\tau)}\bar{\xi} d\tau \\ &= P(\bar{T}) \cdot \bar{\xi} \\ &= \bar{x}. \end{aligned}$$

As it has been mentioned above $\bar{u}(t)$ is the *cheapest* input function in terms of the energy, to drive $\Sigma$ to $\bar{x}$ in time $\bar{T}$. It holds:

$$\|\bar{u}\|^2 = \bar{\xi}^*\ P(\bar{T})\ \bar{\xi}.$$

If the system is completely reachable we have $P(\bar{T})$ positive definite and hence

$$\|\bar{u}\|^2 = \bar{x}^* P(\bar{T})^{-1} \bar{x}.$$

### 1.1.4 Observability

To be able to modify the dynamical behavior of a system, very often the stable $x$ needs to be available. This leads to the question whether a state $x(t)$ can be reconstructed from observing the output $y(\tau)$ on an interval $[T, T+t]$ only. Without loss of generality we assume $T = 0$ and $u(\tau) = 0$ for all $\tau \geq 0$. Therefore, we ignore the input and consider the linear system $\Sigma = \left(\begin{array}{c|c} A & \\ \hline C & \end{array}\right)$ only.

**Definition 1.5.** *Given* $\Sigma = \left(\begin{array}{c|c} A & \\ \hline C & \end{array}\right)$, $A \in \mathbb{R}^{n\times n}$, $C \in \mathbb{R}^{p\times n}$, *then a state* $\bar{x}$ *is "unobservable" if* $y(t) = C\phi(0; \bar{x}, t) = 0 \quad$ *for all* $t \geq 0$.

Then $\bar{x}$ is indistinguishable from the zero state for all $t \geq 0$. The unobservable subspace $\mathcal{N} \subseteq \mathbb{R}^n$ of $\Sigma$ is the set containing all unobservable states. $\Sigma$ is *completely observable* if $\mathcal{N} = \{0\}$. Similar to the reachability matrix $R$ we can deduce an observability matrix $\sigma$ that helps characterizing the unobservable space.

**Theorem 1.2.** *Given* $\Sigma = \left(\begin{array}{c|c} A & \\ \hline C & \end{array}\right)$ *for both discrete- and continuous-time case,* $\mathcal{N}$ *is a linear subspace of* $\mathbb{R}^n$*, given by*

$$\mathcal{N} = \ker\, \sigma(C, A),$$

*with*

$$\sigma = \begin{pmatrix} C \\ CA \\ \vdots \\ CA^{k-1} \\ \vdots \end{pmatrix}.$$

Due to Cayley-Hamilton theorem the kernel of the infinite matrix $\sigma$ is determined at most by the first $n$ terms, i.e., $CA^{k-1}$, $k = 0, \ldots, n-1$. On the other hand according to the reachability gramian, in following we define the finite observability gramian.

**Definition 1.6.** *The finite observability gramian at time* $t < \infty$ *is defined as*

$$\underbrace{Q(t) = \int_0^t e^{A^*\tau} C^* C e^{A\tau} d\tau}_{\text{continuous time systems}} \quad | \quad \underbrace{Q(t) = Q_t^*(C, A) Q_t(C, A)}_{\text{discrete time systems}}.$$

The observability gramian has some valuable properties:

i) $Q(t) = Q^*(t)$ is positive semi-definite,

ii) $\ker Q(t) = \ker \sigma(C, A)$.

The energy of the output function $y$ at time $T$ caused by the state $x$ is denoted by $\|y\|$. In terms of the observability gramian, this energy can be expressed as

$$\|y\|^2 = x^* Q(T) x.$$

### 1.1.5 Infinite gramians

The finite gramians have to be calculated by evaluating an integral. The situation changes, when we consider the infinite gramians:

$$\begin{array}{rcl} P & = & \int_0^\infty e^{A\tau} B B^* e^{A^*\tau} d\tau, \\ Q & = & \int_0^\infty e^{A^*\tau} C^* C e^{A\tau} d\tau, \end{array}$$

that are defined for a stable system $\Sigma = \left(\begin{array}{c|c} A & B \\ \hline C & D \end{array}\right)$, instead. Now $P, Q$ can be determined by solving linear equations, namely *Lyapunov-equations*:

- The continuous Lyapunov equations

$$\begin{array}{rcl} AP + PA^* + BB^* & = & 0, \\ A^*Q + QA + C^*C & = & 0, \end{array} \qquad (1.6)$$

- The discrete Lyapunov equations

$$\begin{array}{rcl} APA^* + BB^* & = & P, \\ A^*QA + C^*C & = & Q. \end{array}$$

We can derive the first equation in (1.6) in the following way

$$\begin{array}{rcl} AP + PA^* & = & \int_0^\infty [Ae^{A\tau} BB^* e^{A^*\tau} + e^{A\tau} BB^* e^{A^*\tau} A^*] d\tau \\ & = & \int_0^\infty d(e^{A\tau} BB^* e^{A^*\tau}) \\ & \overset{A \text{ stable}}{=} & -BB^*. \end{array}$$

**Lemma 1.1.** *Let $P$ and $Q$ denote the infinite gramians of a stable linear system $\Sigma$.*

*a) The "minimal energy required" to steer the state of the system $\Sigma$ from 0 to $x_\tau$ is given by*

$$x_\tau^* P^{-1} x_\tau.$$

*b) The "maximal energy produced" by observing the output of the system whose initial state is $x_0$ is given by*

$$x_0^* Q x_0.$$

The statements in the lemma provide a way to determine states that are hard to reach and/or hard to observe. Suppose that

- $P = X_P \Sigma_P X_P^*$, $X_P = (\bar{x}_{P,1}, \ldots, \bar{x}_{P,n})$ is the singular value decomposition (SVD) of $P$. Let $\bar{u}_{P,r}$ be the optimal input function that steers $\Sigma$ to $\bar{x}_{P,r}$, then $\|\bar{u}_{P,r}\| = \frac{1}{\sigma_{P,r}}$.
- $Q = X_Q \Sigma_Q X_Q^*$, $X_Q = (\bar{x}_{Q,1}, \ldots, \bar{x}_{Q,n})$ is the SVD of $Q$. Then the output energy produced by the state $\bar{x}_{Q,r}$ is $\|\bar{y}_{Q,r}\| = \sigma_{Q,r}$.

This shows that those states are hard to reach that have a significant component in the span of those eigenvectors of $P$ that correspond to small eigenvalues. Those states that have a significant component in the span of the eigenvectors of $Q$ that correspond to small eigenvalues are hard to observe.

The idea is to find a basis transformation $T$, such that the gramians of the transformed system satisfy the following:

$$\bar{P} = \bar{Q} = \begin{pmatrix} \bar{\sigma}_1 & & \\ & \ddots & \\ & & \bar{\sigma}_n \end{pmatrix} \quad (\bar{\sigma}_1 \geq \ldots \geq \bar{\sigma}_n \geq 0)$$

A system with this property is said to be *balanced.* A transformation $T$ of the state space transforms the gramians in the following way:

$$\bar{P} = TPT^*, \quad \bar{Q} = T^{-*}QT^{-1}.$$

However the product turns out to be an *input-output invariant* as:

$$\bar{P}\bar{Q} = T(PQ)T^{-1},$$

i.e., it relates to the original system's product of the gramians by the similarity transformation. The eigenvalues of the product of the reachability and the observability gramians are input-output invariants. They are related to the Hankel singular values which we will introduce in Section 1.1.7.

### 1.1.6 Realization problem

We have seen that the internal representation of a linear system $\Sigma = \left(\begin{array}{c|c} A & B \\ \hline C & D \end{array}\right)$ can be transformed to an external one quite easily.

**Definition 1.7.** *Given the sequence of $p \times m$ matrices $h_k$, $k > 0$ then the "realization problem" consists of finding a positive integer $n$ and constant matrices $(C, A, B)$ such that:*

$$h_k = CA^{k-1}B, \; C \in \mathbb{R}^{p \times n}, \; A \in \mathbb{R}^{n \times n}, \; B \in \mathbb{R}^{n \times m}, \; k \geq 1. \tag{1.7}$$

*The triple $C$, $A$, $B$ is then called a realization of the sequence $h_k$, which is then called "realizable". $C, A, B$ is a "minimal realization" if it is of the smallest dimension $n$ of all possible realizations.*

The Definition 1.7 leads us to the following main questions:

i) Existence,

ii) Uniqueness,

iii) Construction.

The main tool for answering these questions is the *Hankel matrix*:

$$\mathscr{H} = \begin{pmatrix} h_1 & h_2 & h_3 & \dots \\ h_2 & h_3 & \dots & \\ h_3 & \dots & & \\ \vdots & & & \end{pmatrix}.$$

Here we just give the answers, for the proof see [1]:

i) The sequence $\{h_k\}$ ($k \in \mathbb{N}$) is realizable if and only if rank $\mathscr{H} = n < \infty$.

ii) 
   - The dimension of any solution is a least $n$. All realizations that are minimal are reachable and observable and vice versa.
   - All minimal realizations are equivalent.

iii) The Silverman realization algorithm constructs $(C, A, B)$. It is based on finding an $n \times n$ submatrix of $\mathscr{H}$ of rank $n$ and selecting certain rows and columns to construct $C, A, B$.

### 1.1.7 Hankel operator

The *Hankel operator* $\mathscr{H}$ is induced by the convolution operator by restricting its domain and co-domain. In the continuous time case $\mathscr{H}$ is defined by:

$$\mathscr{H} : \mathscr{L}_2^m(\mathbb{R}_-) \to \mathscr{L}_2^p(\mathbb{R}_+), \quad u_- \mapsto y_+$$

with

$$y_+(t) = \mathscr{H}(u_-)(t) = \int_{-\infty}^{0} h(t-\tau)u_-(\tau)d\tau, \quad t \geq 0.$$

Hence, $\mathscr{H}$ is mapping the past inputs to the future outputs.

**Definition 1.8.** *The "Hankel norm" $\|\Sigma\|_H$ is defined as*

$$\|\Sigma\|_H = \sigma_{\max}(\mathscr{H}),$$

*where $\sigma_{\max}(\mathscr{H})$ is the largest singular value of $\mathscr{H}$.*

$\|\Sigma\|_H$ is the induced norm in the frequency domain space $\mathscr{L}_2(i\mathbb{R})$ of $\mathscr{H}$. A relation between the gramians $P, Q$ and the Hankel operator is given by the Lemma 1.2, for more details see [1].

**Lemma 1.2.** *Given a reachable, observable and stable system $\Sigma$ of dimension $n$. The Hankel singular values of $\Sigma$ are equal to the positive square roots of the eigenvalues of the product of gramians $PQ$:*

$$\sigma_k(\Sigma) = \sqrt{\lambda_k(PQ)}, \quad k = 1, \ldots, n.$$

## 1.2 Differential Algebraic Equations

In this part, we will first briefly discuss modified nodal analysis (MNA) [36] and basic types of DAEs. In the next part we will study different index concepts for DAEs which gives an indication on the complexity of the problem of solving a particular DAE.

The mathematical description of an integrated circuit is mostly based, in current industrial environment, on a network representation of the entire circuit and on a lumped description of each single electron device [12]. Many modeling paradigms can be then derived combining *Kirchhoff's current law* (KCL), *Kirchhoff's voltage law* (KVL) and *elemental constitutive relations* [13, 16, 34]. However, one looks among these many paradigms established over the others for the possibility it gives to automate the assembly of circuit equations in a numerical framework while maintaining reasonably low the number of unknowns: modified nodal analysis (MNA) [36].

In its original formulation MNA keeps the node-potential vector $e$, the voltage-source current vector $i_V$ and the inductor current vector $i_L$ as unknown quantities and derives a closed system of equations enforcing KCL at each node of the circuit, expressing the current through each voltage-controllable element in terms of node-potentials and complementing the system with the constitutive relations for current-controllable elements. In general a set of *differential-algebraic* equations will stem from this procedure [32,59].

From the practical point of view the most interesting feature of MNA is the possibility to assemble the whole set of equations describing a circuit on the base of elemental contributions. This characteristic, stemming from KCL, is inherent in the derivation of MNA and remains even at the discrete level. In fact the structure most commonly adopted in the design of a software package for transient circuit simulation has at its core a set of element evaluators that provides a non-linear solver with the local contribution to the overall Jacobian matrix and the residual. It is precisely these contributions that can be reduced via MOR to improve the simulation efficiency.

The set of equations derived from MNA is a DAE. More specifically, it is a *linear implicit (quasilinear)* DAE (1.8) or even linear form (1.10). To set up algorithms that solve these network equations numerically a transformation into semi-explicit form (1.11) proves to be useful, see [35]. Here the equations can be written in the following form:

$$f(\dot{x}, x, t) := A(x,t)\dot{x}(t) + g(x,t) = 0, \tag{1.8}$$

with $A(x,t) \in \mathbb{R}^{n\times n}$, $f : \mathbb{R}^n \times \mathbb{R}^n \times I \to \mathbb{R}^n$ and $g : \mathbb{R}^n \times I \to \mathbb{R}^n$. In terms of $q(x,t)$ and $j(x,t)$ we have:

$$f(\dot{x}, x, t) := \frac{\partial q(x,t)}{\partial x}\dot{x}(t) + \frac{\partial q(x,t)}{\partial t} + j(x,t) = 0, \tag{1.9}$$

where $A(x,t) = \frac{\partial q(x,t)}{\partial x}$ and $g(x,t) = \frac{\partial q(x,t)}{\partial t} + j(x,t)$. This form is called quasilinear, because the equations are linear in $\dot{x}(t)$.

Equations derived from a sample network could be in the following form:

$$A\dot{x} + Bx + s(t) = 0, \tag{1.10}$$

with constant coefficient matrix $A,\ B \in \mathbb{R}^{n\times n}$. An example of this type is:

$$\begin{bmatrix} 1 & 0 \\ 0 & 0 \end{bmatrix} \begin{bmatrix} \dot{x}_1 \\ \dot{x}_2 \end{bmatrix} + \begin{bmatrix} 0 & 1 \\ 1 & 0 \end{bmatrix} \begin{bmatrix} x_1 \\ x_2 \end{bmatrix} = \begin{bmatrix} 0 \\ t \end{bmatrix},$$

which is the same as

$$\begin{aligned} \dot{x}_1 + x_2 &= 0, \\ x_1 &= t. \end{aligned}$$

If the matrices $A$ and $B$ do not depend on $x$, but do depends on $t$, we are dealing with linear time-varying DAEs:

$$A(t)\dot{x}(t) + B(t)x(t) = s(t).$$

Such equations are studied for analyzing small perturbations around periodic steady-state solutions which have applications in RF circuit design. Sometimes the equations can be separated into differential equations and into algebraic equations by a suitable reordering of the components of the unknown vector $x$ and of the equations themselves. More precisely:

$$x(t) = \begin{bmatrix} y(t) \\ z(t) \end{bmatrix}$$

and the DAE system reads

$$\begin{aligned} \dot{y}(t) &= f(y, z, t), \\ 0 &= g(y, z, t). \end{aligned} \tag{1.11}$$

As we mentioned above, linear implicit DAEs can be written as

$$A(x(t), t)\dot{x}(t) + g(x(t), t) = 0. \tag{1.12}$$

Note that if the matrix $A(x(t), t)$ is nonsingular, (1.12) can be transformed into an ODE by multiplying with $A^{-1}$.

## Index of DAEs

To distinguish the degree in solving DAEs we associate an index. Index is a notion used in the theory of DAEs for measuring the distance from a DAE to its related ODE. The index is a nonnegative integer that provides useful information about the mathematical structure and potential complications in the analysis and the numerical solution of the DAE. In general, the higher the index of a DAE, the more difficulties one can expect for its numerical solution. There are various ways to measure the complexity of DAEs. The most important ones are:

- Differential index,
- Tractability index,
- Perturbation index,
- Geometrical index.

In the nonlinear case they are equivalent, see [59, 67, 68]. Howerver the genralizations to the nonllinear case are different.

### Differential index

The differential index, see [27,28,35,63], is based on taking derivatives of the DAE and turning it into an explicit ODE. It counts the differentiations that are necessary.

**Definition 1.9.** *The differential index $k$ of a nonlinear, sufficiently smooth DAE of the form* (1.9) *is the smallest $k$ such that the system:*

$$\begin{aligned} f(\dot{x},x,t) &= 0, \\ \tfrac{d}{dt} f(\dot{x},x,t) &= 0, \\ &\vdots \\ \tfrac{d^k}{dt^k} f(\dot{x},x,t) &= 0, \end{aligned} \tag{1.13}$$

*such that the set of equations (1.13) allows to extract an explicit system of ODEs $\dot{x} = \varphi(x,t)$ by algebraic manipulations. This system is called the " underlying ODE".*

We use equation (1.14), which is derived by MNA from a linear circuit [2], to explain the differential index:

$$f(\dot{x},x,t) = \begin{bmatrix} C & 0 \\ 0 & 0 \end{bmatrix} \dot{x} + \begin{bmatrix} 0 & -1 \\ 1 & 0 \end{bmatrix} x - \begin{bmatrix} 0 \\ V(t) \end{bmatrix} = 0, \tag{1.14}$$

where the vector of unknowns $x$ is $(v_{n_1}(t), i_{b_1}(t))^T$. These equations can be rewritten as:

$$C\tfrac{d}{dt}(v_{n_1}(t)) - i_{b_1}(t) = 0, \tag{1.15a}$$

$$v_{n_1}(t) \quad = \quad V(t). \tag{1.15b}$$

Differentiating (1.15b) gives:

$$\frac{d}{dt}v_{n_1}(t) = \frac{d}{dt}V(t). \tag{1.16}$$

Multiplying equation (1.16) by $C$ ad substituting the result from (1.15a) gives:

$$-i_{b_1}(t) = -C\frac{d}{dt}V(t).$$

Differentiating and reordering the terms gives us the following ODE

$$\frac{d}{dt}i_{b_1}(t) = C\frac{d^2}{dt^2}V(t).$$

Thus the network in our example leads to a DAE of differential index 2. Note that the resulting ordinary differential equation contains the second derivative of the input signal. We will apply to the concept of differential index when we introduce the *Kronecker canonical form* in the next chapter.

### Tractability index

The tractability index, see [63, 67, 68], provides a decomposition of the DAE into its regular ODE as well as algebraic equations and differentiation problems. It is suitable for a detailed analysis of DAEs and requires only minimal smoothness conditions on the function $f$. In order to explain the tractability index, we need the following definitions.

**Definition 1.10.** *A linear operator $Q \in L(\mathbb{R}^m)$ is called a projector if the relation $Q^2 = Q$ is fulfilled.*[1]

An example of a projector is

$$Q = \begin{bmatrix} 1 & 0 \\ 3 & 0 \end{bmatrix}.$$

We remark that a projector along a subspace is not unique. A projector is only unique when it is along a subspace and onto some other subspace. For the definition of the tractability index we need the following matrix chain

$$\begin{aligned}
A_0(\dot{x}, x, t) &= f_{\dot{x}}(\dot{x}, x, t), \\
B_0(\dot{x}, x, t) &= f_x(\dot{x}, x, t) - A_0(\dot{x}, x, t)\dot{P}_0(t), \\
A_{i+1}(\dot{x}, x, t) &= A_i(\dot{x}, x, t) + B_i(\dot{x}, x, t)Q_i(\dot{x}, x, t), \\
B_{i+1}(\dot{x}, x, t) &= B_i(\dot{x}, x, t)P_i(\dot{x}, x, t) \\
&\quad -A_{i+1}(\dot{x}, x, t)\tfrac{d}{dt}(P_0(t)\dots P_{i+1}(\dot{x}, x, t))P_0(t)\dots P_i(\dot{x}, x, t),
\end{aligned} \tag{1.17}$$

[1] We call a projector $Q \in L(\mathbb{R}^m)$ a projector onto a subspace $S \subseteq \mathbb{R}^m$ if $\mathrm{im}[Q] = S$ and a projector $Q \in L(\mathbb{R}^m)$ a projector along a subspace $S \subseteq \mathbb{R}^m$ if $\ker[Q] = S$.

for $i \geqslant 0$. Here, $Q_i(\dot{x}, x, t)$ denotes a continuously differentiable projector onto $\ker[A_i(\dot{x}, x, t)]$ and $P_i(\dot{x}, x, t) = I - Q_i(\dot{x}, x, t)$ a projector along $\ker[A_i(\dot{x}, x, t)]$. We assume that $\ker[A_i(\dot{x}, x, t)]$ only depends on $t$ and thus $P_0(\dot{x}, x, t) = P_0(t)$.

**Definition 1.11.** *The nonlinear DAE (1.9) is called index $k$ tractable if the matrices $A_i(\dot{x}, x, t)$ in (1.17) are singular and of constant rank on $G_f := \mathbb{R}^n \times \mathbb{R}^n \times I$ for $0 \leq i \leq k-1$ and $A_k(\dot{x}, x, t)$ remains nonsingular on $G_f$.*

We use the same equation (1.14) to compute the tractability index of the DAE. The equation we find can be rewritten as linear DAE with constant coefficients of the form (1.10):

$$A\dot{x} + Bx(t) = s(t),$$

where $x$ represents the vector of unknowns, $A, B$ are constant matrices and $s(t)$ a vector function. Thus, in this example

$$A = \begin{bmatrix} C & 0 \\ 0 & 0 \end{bmatrix}, \quad B = \begin{bmatrix} 0 & -1 \\ 1 & 0 \end{bmatrix}, \quad \text{and } s(t) = \begin{bmatrix} 0 \\ V(t) \end{bmatrix}.$$

The matrices $A_0$ and $B_0$ are the same as $A$ and $B$, respectively. Because the DAE is linear with constant coefficients, a projector on ker[A] is also a constant matrix. We choose the projector

$$Q_0 = \begin{bmatrix} 0 & 0 \\ 0 & 1 \end{bmatrix}.$$

Thus we get

$$A_1 = \begin{bmatrix} C & -1 \\ 0 & 0 \end{bmatrix}, B_1 = \begin{bmatrix} 0 & 0 \\ 1 & 0 \end{bmatrix}.$$

Since $A_1$ is singular, we next have to compute a projector $Q_1$ onto $\ker[A_1]$. It is given by

$$Q_1 = \begin{bmatrix} 1 & 0 \\ C & 0 \end{bmatrix},$$

which leads to

$$A_2 = \begin{bmatrix} C & -1 \\ 1 & 0 \end{bmatrix}.$$

Since $A_2$ is regular, we can conclude that the DAE has tractability index-2. We are able to split $x$ into components

$$\begin{aligned} x &= Q_0x + P_0x \\ &= Q_0x + P_0Q_1x + P_0P_1x. \end{aligned}$$

In general, $P_0P_1x$ corresponds to the part for which a pure differential equation is given. In our example $P_0P_1x = 0$ and thus there are no variables for which a pure differential equation is given. $P_0Q_1x = (v_{n_1}(t), 0)^T$, hence for the variable $v_{n_1}(t)$ there is a pure algebraic relation. We speak of an index-1 variable in this case. $Q_0x$ corresponds to $(0, i_{b_1}(t))^T$. For $i_{b_1}(t)$ we have to differentiate the last equation once. This kind of variable is called an index-2 variable. The differential index and the tractability index are the same for our example. This is not just coincidence. The DAE of this example is a linear DAE with constant coefficients. For this type see [68].

**Lemma 1.3.** *A linear DAE has the tractability index $k$ if and only if it has the differential index $k$.*

Differential index and tractability index are mostly used in literature. For completeness, we also give the definition of the perturbation index and of the geometrical index.

### Perturbation index

The perturbation index measures the sensitivity of solutions with respect to perturbations of the given problem, see [35, 63].

**Definition 1.12.** *The DAE (1.9) has the perturbation index $k > 0$ along a solution $x(t)$ on $I_0 = [0, T]$ if $k$ is the smallest integer such that, for all functions $\hat{x}(t)$ having a defect*

$$f(\dot{\hat{x}}(t), \hat{x}(t), t) = \delta(t),$$

*there exists an estimate*

$$\|x(t) - \hat{x}(t)\| \leq C\left(\|x(0) - \hat{x}(0)\| + \max_{0\leq t\leq T}\|\delta(t)\| + \max_{0\leq t\leq T}\|\dot{\delta}(t)\| + \cdots + \max_{0\leq t\leq T}\|\delta^{(k-1)}(t)\|\right)$$

*for a constant $C > 0$ whenever the expression on the right-hand side is sufficiently small.*

Note that the perturbation index concept requires information about the solution of the DAE. Again an example for illustration [35]. Consider the semi-explicit system of equations (1.11) with regular partial derivative $g_z$ near the solution. Then the perturbed system is

$$\begin{aligned} \dot{\hat{y}} &= f(\hat{y}, \hat{z}) + \delta(t), \\ 0 &= g(\hat{y}, \hat{z}) + \eta(t). \end{aligned}$$

Since $g_z$ is invertible by hypothesis, the difference $\hat{z} - z$ can be estimated with the help of the implicit function theorem without any differentiation to

$$\|\hat{z}(t) - z(t)\| \leq C_1 \left(\|\hat{y}(t) - y(t)\| + \|\eta(t)\|\right).$$

Using this estimate, a Lipschitz condition for $f$ and Gronwall's Lemma we end up with:

$$\|\hat{y}(t) - y(t)\| \leq C\left(\|\hat{y}(0) - y(0)\| + \max_{0\leq t\leq T}\|\delta(t)\| + \max_{0\leq t\leq T}\|\eta(t)\|\right)$$

for all $t$ in a bounded interval $[0, T]$ and we conclude that the perturbation index is 1. It was believed for some time that the differential index and the perturbation index are differing at most by one which is not true, for a counter example, see [35].

### Geometrical index

The geometrical index [53] provides useful insights into the geometrical and analytical nature of DAEs. For this approach, DAEs are regarded as explicit ODEs on manifolds. We want to illustrate this idea with an example. We use the equations (1.14). The only change is that the voltage source is a constant 5V source. This results in the following system of equations

$$C(\dot{x}_1) - x_2 = 0, \tag{1.19a}$$

$$x_1 = 5. \tag{1.19b}$$

It is clear that all solutions of the DAE (1.19a)-(1.19b) lie on the manifold

$$\Gamma_1 = \left\{ (x_1, x_2)^T : x_1 = 5 \right\}.$$

This shows that we cannot find a solution if the starting point does not belong to this manifold. But this is not the only restriction for the solution. If we differentiate the equation (1.19b) and put it into equation (1.19a) we see that a solution of the system (1.19a)-(1.19b) belongs also to the manifold

$$\Gamma_2 = \left\{ (x_1, x_2)^T : x_2 = 0 \right\}.$$

The solution of the DAE (1.19a)-(1.19b) may be interpreted as an explicit ODE (1.19a) on the manifold $\Gamma_1 \cap \Gamma_2$. Namely:

$$C(\dot{x}_1) - x_2 = 0 \quad \stackrel{x_2=0}{\Longrightarrow} \quad x_1 = 5.$$

The reason we changed the voltage source into a constant source is that the geometrical index is based on autonomous problems, i.e. problems of the form

$$f(\dot{x}, x) = 0. \tag{1.20}$$

In circuit simulations, autonomous problems arise, when dealing with free running oscillators. We sketch the idea of the geometrical approach. Assume the zero set $M_0 = f[\{0\}]$ is a smooth manifold of $T\mathbb{R}^m = \mathbb{R}^m \times \mathbb{R}^m$. Then the DAE (1.20) may be written in the form

$$(x, \dot{x}) \in M_0.$$

In our example $M_0$ is

$$M_0 = \left\{ (x_1, x_2, p_{x_1}, p_{x_2})^T : x_1 - 5 = 0, Cp_{x_1} - x_2 = 0 \right\}.$$

That in turn implies that any solution $x = x(t)$ of (1.20) has to satisfy $x(t) \in W_0 = \pi(M_0)$, where $\pi : T\mathbb{R}^m \to \mathbb{R}^m$ is the canonical projection onto the first component. Thus

$$W_0 = \left\{ (x_1, x_2)^T : x_1 - 5 = 0 \right\}.$$

In general, if $W_0$ is a sub-manifold of $\mathbb{R}^m$, $(x(t), \dot{x}(t))$ belongs to the tangent bundle $TW_0$ of $W_0$. In other words, our solution belongs to

$$\begin{aligned} M_1 &= M_0 \cap TW_0 \\ &= \left\{ (x_1, x_2, p_{x_1}, p_{x_2})^T : x_1 - 5 = 0, Cp_{x_1} - x_2 = 0, p_{x_1} = 0 \right\}. \end{aligned}$$

This is a reduction to (1.20). We can reduce even more, using that $W_i = \pi(M_i)$. In our example we get

$$\begin{aligned} M_2 &= M_1 \cap TW_1 \\ &= \left\{ (x_1, x_2, p_{x_1}, p_{x_2})^T : x_1 - 5 = 0, Cp_{x_1} - x_2 = 0, p_{x_1} = 0, p_{x_2} = 0 \right\}. \end{aligned}$$

Thus $W_2 = W_1$ and $M_3 = M_2$. We are now ready to state the definition of the geometrical index.

**Definition 1.13.** *The first integer $k$ such that $M_k = M_{k+1}$ holds is called the "geometrical index" of* (1.20).

Our example leads to a geometrical index of 2. This definition 1.13 was proposed without results concerning the existence of manifolds and their properties were given. In [53], a comprehensive analysis of these questions is given. It turns out that the global approach (sketched above) must be replaced by a local one.

# Chapter 2

# Linear model order reduction

This chapter is dealing with the linear systems and linear methods for reduction. Among all the methods two fundamental categories are discussed in details. The first one is Krylov based methods such as passive reduced order interconnect macromodeling algorithm (PRIMA [47]) and structure preserving reduced order interconnect macromodeling (SPRIM [22]) and the second one is balancing truncation based on Hankel singular value approximation such as truncated balance realization (TBR [38, 46, 52]) and poor man's TBR (PMTBR [48]). The concepts of passivity and structural preserving are also studied briefly. The new techniques for reducing semi-explicit system of DAEs are introduced which in the following are extended to all linear DAEs. Finally we briefly review the existing methods for parametric reduction. Among those methods, we studied parameterized interconnect macromodeling via a two-directional Arnoldi process (PIMTAB) in more details.

## 2.1 Linear reduction of dynamical systems

A continuous time-invariant (lumped) multi-input multi-output linear dynamical system can be derived from an RLC circuit by applying modified nodal analysis (MNA) [31]. The system is of the form:

$$\begin{cases} C\frac{dx(t)}{dt} &= -Gx(t) + Bu(t) \\ y(t) &= Lx(t) + Du(t), \end{cases} \tag{2.1}$$

with initial condition $x(0) = x_0$. Here $t$ is the time variable, $x(t) \in \mathbb{R}^n$ is referred as inner state (and the corresponding $n$-dimensional space is called state space), $u(t) \in \mathbb{R}^m$ is an input, $y(t) \in \mathbb{R}^p$ is an output. The dimensionality $n$ of the state vector is called the order of the system. $m$ is the number of inputs and $p$ is the number of outputs and $G \in \mathbb{R}^{n \times n}$ , $B \in \mathbb{R}^{n \times m}$ , $L \in \mathbb{R}^{p \times n}$ , $C \in \mathbb{R}^{n \times n}$, $D \in \mathbb{R}^{p \times m}$ are the state space matrices. Without loss of generality we assume $D = 0$. Moreover, we assume that $C, G,$ and $B$

exhibit the block structure [1]:

$$C = \begin{bmatrix} C_1 & 0 \\ 0 & C_2 \end{bmatrix}, \quad G = \begin{bmatrix} G_1 & G_2 \\ -G_2^T & 0 \end{bmatrix}, \quad B = \begin{bmatrix} B_1 \\ 0 \end{bmatrix}, \tag{2.2}$$

where the sub-blocks $C_1$, $G_1$, and $B_1$ have the same number of rows, and

$$C_1 \geq 0, \; G_1 \geq 0, \; C_2 > 0. \tag{2.3}$$

Here, $\star \geq 0$ or $\star > 0$ means that $\star$ is symmetric and positive semi-definite or positive definite, respectively. The condition (2.3) implies that the matrices $C$ and $G$ satisfy:

$$G + G^T \geq 0 \text{ and } C \geq 0.$$

The matrices $C$ and $G$ in (2.1) are allowed to be singular, and we only assume that the pencil $G + sC$ is *regular*, i.e., the matrix $G + sC$ is singular only for a finite number of values $s \in \mathbb{C}$. For more details on uniqueness of a solution see [35].

The linear system of the form (2.1) is often referred to as the representation of the system in the time domain, or in the state space. Equivalently, one can also represent the system in the frequency domain via the *Laplace* transform. Recall that for a vector-valued function $f(t)$, the Laplace transform $f(t)$ is defined component-wise by

$$F(s) := \mathbb{L}\{f(t)\} = \int_0^\infty f(t)e^{-st}dt, \qquad s \in \mathbb{C}. \tag{2.4}$$

The physically meaningful values of the complex variable $s$ are $s = i\omega$, where $\omega \geq 0$ is referred to as the (angular) frequency. Taking the Laplace transformation of the system (2.1), we obtain the following frequency domain formulation

$$\begin{cases} sCX(s) &= -GX(s) + BU(s) \\ Y(s) &= LX(s), \end{cases} \tag{2.5}$$

where $X(s)$, $Y(s)$ and $U(s)$ represents the Laplace transform of $x(t)$, $y(t)$ and $u(t)$, respectively. For simplicity, we assume that we have zero initial conditions $x(0) = x_0 = 0$ and $u(0) = 0$. Eliminating the variable $X(s)$ in (2.5), we see that the input $U(s)$ and the output $Y(s)$ in the frequency domain are related by the following $p \times m$ matrix-valued rational function

$$H(s) = L \cdot (G + s \cdot C)^{-1} \cdot B. \tag{2.6}$$

$H(s)$ is known as the *transfer function* or *Laplace-domain impulse response* of the linear system (2.1). The following types of analysis are typically performed for a given linear dynamical system of the form (2.1):

[1] If we apply the standard MNA [31].

- Static (DC) analysis, to find out the point to which the system settles in an equilibrium or rest condition, namely $\dot{x}(t) = 0$;
- Steady-state analysis, also called frequency response analysis, to determine the frequency responses $H(s)$ of the system to external steady-state oscillatory (i.e., sinusoidal) excitation;
- Modal frequency analysis, to find the system's natural vibrating frequency modes and their corresponding modal shapes;
- Transient analysis, to compute the output behavior $y(t)$ subject to time-varying excitation $u(t)$;
- Sensitivity analysis, to determine the proportional changes of the time response $y(t)$ and/or steady state response $H(s)$ to a proportional change in system parameters.

## 2.2 MOR Methods

Up to now, the largest group of MOR algorithms applies to linear systems or more precisely linear time-invariant (LTI) systems. Within the MOR algorithms for LTI systems one may distinguish the following most popular classes:

I) MOR algorithms based on Krylov subspace methods.

II) Methods based on Hankel norm approximations and TBR.

Both of those methods apply the concept of approximating a certain part of a high-dimensional state space of the original system by a lower-dimensional reduced space, or in other words, they perform a projection of the original state space. The projections can be constructed in a number of different ways. Basically, MOR techniques aim to derive a system:

$$\begin{cases} \tilde{C}\frac{d\tilde{x}(t)}{dt} & = & -\tilde{G}\tilde{x}(t) + \tilde{B}u(t), \quad \tilde{x}(t) \in \mathbb{R}^q \\ \tilde{y}(t) & = & \tilde{L}\tilde{x}(t) + \tilde{D}u(t), \quad \tilde{x}(0) = \tilde{x}_0, \quad \tilde{y}(t) \in \mathbb{R}^p, \end{cases} \tag{2.7}$$

of order $q$ with $q \ll n$ that can replace the original high-order system (2.1) in the sense, that the input-output behavior, described by the transfer function (2.6) in the frequency domain, of both systems nearly agrees. A common way is to identify a subspace of dimension $q \ll n$, that captures the dominant information of the dynamics and project (2.1) onto this subspace, spanned by some basis vectors $\{v_1, \ldots, v_q\}$.

The reduction can be carried out by means of different techniques. Approaches like SPRIM [22], PRIMA [47], TBR [46,49] and PMTBR [48] project the full problem (2.1) onto a subspace of dimension $q$ that captures the dominant information. The first two rely on Krylov subspace methods. The truncated balancing realization is based on Hankel norm approximation and system information. The latter one exploits the direct relation between the multi-point rational projection framework and the TBR. The PMTBR can take advantage of some a-priori knowledge of the system properties, and is based on a statistical interpretation of the system gramians.

### 2.2.1 Introduction to Krylov Projection Techniques

In recent years, MOR techniques based on Krylov subspaces have become the methods of choice for generating macromodels of large multi-port RLC circuits. Krylov subspace methods provide numerically robust algorithms for generating a basis of the reduced space, such that a certain number of moments of the transfer function of the original system is matched. Consequently, the transfer function of the reduced system approximates the original transfer functions around a specified frequency, or a collection of frequency points [29]. Owing to their robustness and low computational cost, Krylov subspace algorithms proved suitable for the reduction of large-scale systems, and gained considerable popularity, especially in electrical engineering. A number of Krylov-based MOR algorithms have been developed, including techniques based on the Lanczos method [18,26] and the Arnoldi algorithm [47,58,74]. The main drawbacks of these methods are, in general, lack of provable error bounds for the extracted reduced models, and no guarantee for preserving stability and passivity. Nevertheless, it has been demonstrated that if the original system has a specific structure, both stability and passivity can be preserved in the reduced system, by exploiting the fact that congruence transformations preserve the definiteness of a matrix. PRIMA [47] combines the moment matching approach with projection to arrive at a reduced system of type (2.7). Its main feature is that it produces passive reduced models.

#### PRIMA: Passive Reduced-Order Interconnect Macromodeling Algorithm

We consider the system (2.1). Krylov techniques are implicit moment matching approaches, i.e. they are based on the construction of the transfer function in the frequency domain. Let us define the matrices

$$\begin{aligned} A &= -G^{-1} \cdot C, \\ R &= G^{-1} \cdot B. \end{aligned} \tag{2.8}$$

From the expression in (2.1) with unit impulses at the ports, the Laplace transform can be applied to obtain the transfer function. For the details see Section 2.1. Using (2.8), the transfer function can be reformulated as

$$H(s) = L \cdot (I - s \cdot A)^{-1} \cdot R$$

where $I = \mathbb{R}^{n\times n}$ is the identity matrix. The block moments of the transfer function can be defined as the coefficients of the Taylor series expansion around $s = 0$

$$H(s) = M_0 + M_1 \cdot s + M_2 \cdot s^2 + \cdots ,$$

where $M_i = \mathbb{R}^{n\times n}$ are the block moments that can be computed by

$$M_i = L \cdot A^i \cdot R.$$

The block Krylov subspace generated by matrices $A \in \mathbb{R}^{n\times n}$ and $R = [r_1 r_2 \ldots r_m] \in \mathbb{R}^{n\times m}$ is defined as

$$\mathrm{Kr}(A, R, q) = \mathrm{colspan}[R, A \cdot R, A^2 \cdot R, \ldots, A^{k-1} \cdot R, A^k \cdot r_1, \ldots, A^k \cdot r_l],$$

where $k = \left\lfloor \frac{q}{m} \right\rfloor, l = q - k \cdot m$.

This means that the Krylov subspace spans the combination of moment vectors generated by the different sources in the circuit. So any basis of this subspace can be used to project the system matrices onto it. Then the $k$ moments of the original transfer function are matched by the projected system, i.e. the reduced system. The Krylov subspace can be efficiently generated via robust and well developed algorithms such as the block Arnoldi algorithm or the Lanczos process. The most popular Krylov projection algorithm is the PRIMA algorithm, where an orthonormal matrix appears in the form:

$$V = (v_1, \ldots, v_q) \in \mathbb{R}^{n\times q}. \tag{2.9}$$

The columns of $V$ in (2.9) span the Krylov subspace of order $q$. It is built and applied via a congruence transformation over the system matrices for obtaining a reduced order model of size $q$ that matches the first $k$ block moments of the original transfer function. This projection is performed in the following way

$$\begin{aligned} \tilde{C} &= V^T \cdot C \cdot V \\ \tilde{G} &= V^T \cdot G \cdot V \\ \tilde{B} &= V^T \cdot B \\ \tilde{L} &= L \cdot V \ , \end{aligned} \tag{2.10}$$

where $\tilde{G} \in \mathbb{R}^{q\times q}$ , $\tilde{B} \in \mathbb{R}^{q\times m}$ , $\tilde{L} \in \mathbb{R}^{p\times q}$ , $\tilde{C} \in \mathbb{R}^{q\times q}$ are the reduced order system matrices and $V \in \mathbb{R}^{n\times q}$ is the projector, i.e. the orthonormal basis for the Krylov subspace. Finally, the reduced system can be expressed via its transfer function:

$$\tilde{H}(s) = \tilde{L} \cdot (\tilde{G} + s \cdot \tilde{C})^{-1} \cdot \tilde{B}. \tag{2.11}$$

The PRIMA or in general any algorithm that applies congruence projection shows another advantage. Necessary and sufficient conditions for the square ($m = p$) system (2.1) to be passive is that the transfer function in (2.6) is positive real, which means that:

- $H(s)$ is analytic for $\mathrm{Re}(s) > 0$;
- $H(\bar{s}) = \overline{H(s)}$, for all $s \in \mathbb{C}$;
- $H(s) + (H(s))^* \geq 0$ for $\mathrm{Re(s)} > 0$ where $(H(s))^* = B^*(G^* + \bar{s}C^*)^{-1}L^*$.

The second condition is satisfied for real systems and the third condition implies the existence of a rational function with a stable inverse. Any congruence transformation applied to the system matrices satisfies the previous conditions if the original system satisfies them, and so preserves the passivity of the system if the following conditions are true:

- The system matrices are positive definite, $C, G \geq 0$,
- $B = L^T$.

These conditions are sufficient, but not necessary. They are usually satisfied in the case of electrical circuits, which makes congruence-based projection methods very popular in circuit simulations.
A drawback of Krylov-based projection techniques are the lack of efficient techniques for error control. Error estimators do exist but are seldom used in practice as they are expensive and cumbersome to use. However, by the used sparse matrix techniques, the methods are very efficient and generally produce pretty good results, which has led to their widespread usage in very large scale integration (VLSI) settings where reduction of large passive interconnect systems is often required.

### SPRIM: Structure-Preserving Reduced-Order Interconnect Macromodeling

We discussed some of the advantages and disadvantages of Krylov-based techniques, especially PRIMA in the previous part. However, PRIMA does not preserve the structure of the system matrices which is of an interest when trying to realize the reduced model. SPRIM [22], an adoption of this method, preserves block structures of the circuit matrices and generates provably passive and reciprocal [1] macromodels of multi-port RLC circuits. The SPRIM models match twice as many moments as the corresponding PRIMA models obtained with the same amount of computational work. Also SPRIM is less restrictive to matrices $C, G$ in system (2.1), see [24].

Recall PRIMA combines projection with block Krylov subspaces, see PRIMA algorithm in [47]. More precisely, the reduced transfer function (2.11) where the matrix $V$ from (2.9), is chosen such that its columns span the $k$th block Krylov subspace $\mathrm{Kr}(A, R, q)$, i.e.,

$$\mathrm{span}V = \mathrm{Kr}(A, R, q). \tag{2.12}$$

[1] A two-terminal element is said to be reciprocal, if a variation of the values of one terminal immediately has the reverse effect on the other terminal's value. Linear characteristics obviously have this property.

**Theorem 2.1.** *Let $k = m_1 + m_2 + \cdots + m_q$ and the matrix $V$ in (2.9) satisfying (2.12). Then, the first $q$ moments in the expansions of $H(s)$ and of $\tilde{H}(s)$ around the expansion point $s_0$ are identical:*

$$H(s) = \tilde{H}(s) + O((s - s_0)^q).$$

Recall that the matrices $G$, $C$ and $B$ in (2.6) exhibit the particular block structure (2.2). However, for the PRIMA reduced-order model, the reduced matrices in (2.10) cannot preserve the block structure (2.2). Now, let $\tilde{V}$ be any matrix, possibly with more than $n$ columns, such that the space spanned by the columns of $\tilde{V}$ contains the $k$th block Krylov subspace $\mathrm{Kr(A,R)}$, i.e.,

$$\mathrm{Kr(A,R)} \subseteq \mathrm{span}\tilde{\mathrm{V}}. \tag{2.13}$$

Using such a matrix $\tilde{V}$, we define a reduced-order model as follows:

$$\hat{H}(s) = \hat{L} \cdot (\hat{G} + s \cdot \hat{C})^{-1} \cdot \hat{B},$$

where

$$\begin{aligned} \hat{C} &= \tilde{V}^T \cdot C \cdot \tilde{V} \\ \hat{G} &= \tilde{V}^T \cdot G \cdot \tilde{V} \\ \hat{B} &= \tilde{V}^T \cdot B \\ \hat{L} &= L \cdot \tilde{V}. \end{aligned} \tag{2.14}$$

Like the Theorem 2.1, we have the following generalizing result.

**Theorem 2.2.** *Let $k = m_1 + m_2 + \cdots + m_q$ and $\tilde{V}$ be a matrix, possibly with more than $n$ columns, such that (2.13) is satisfied. Then, the first $q$ moments in the expansions of $H(s)$ and of $\hat{H}(s)$ (defined by (2.13) and (2.14)) around $s_0$ are identical:*

$$H(s) = \hat{H}(s) + O((s - s_0)^q).$$

For the proof see [23]. In following we will introduce the steps of SPRIM algorithm.

1. The matrices in (2.2) are chosen as an input and their sub-blocks satisfy the condition (2.3), and the expansion point is $s_0$.
2. $A$ and $R$ are defined as in (2.8).
3. Construct the basis matrix $V = (v_1, \ldots, v_q)$ which is the outcome of running any favorite block Krylov subspace methods (applied to $A$ and $R$):

$$\mathrm{span}V = \mathrm{Kr(A,R,q)}.$$

4. Let $V = \begin{bmatrix} V_1 \\ V_2 \end{bmatrix}$ be the partitioning of $V$ corresponding to the block structure of $G$ and $C$.

5. Set

$$\begin{array}{ll} \tilde{C}_1 = V_1^T \cdot C_1 \cdot V_1, & \tilde{C}_2 = V_2^T \cdot C_2 \cdot V_2, \\ \tilde{G}_1 = V_1^T \cdot G_1 \cdot V_1, & \tilde{G}_2 = V_1^T \cdot G_2 \cdot V_2, \\ \tilde{B}_1 = V_1^T \cdot B_1, & \tilde{L}_1 = L_1 \cdot V_1. \end{array}$$

and

$$\tilde{C} = \begin{bmatrix} \tilde{C}_1 & 0 \\ 0 & \tilde{C}_2 \end{bmatrix}, \quad \tilde{G} = \begin{bmatrix} \tilde{G}_1 & \tilde{G}_2 \\ -\tilde{G}_2^T & 0 \end{bmatrix}, \quad \tilde{B} = \begin{bmatrix} \tilde{B}_1 \\ 0 \end{bmatrix}.$$

6. The transfer function in the first-order form of (2.11) can be defined and in second-order form

$$\tilde{H}(s) = \tilde{L}_1 \cdot (\tilde{G}_1 + s \cdot \tilde{C}_1 + \tfrac{1}{s} \cdot \tilde{G}_2^T \tilde{C}_2^{-1} \tilde{G}_2)^{-1} \cdot \tilde{B}_1. \tag{2.15}$$

### Properties of SPRIM

Theorem 2.2 implies that the SPRIM has at least the same number of moments as PRIMA but the SPRIM models match twice as many moments as in the corresponding PRIMA models. We define the SPRIM projection $\tilde{V}$ as follows:

$$\tilde{V} = \begin{bmatrix} V_1 & 0 \\ 0 & V_2 \end{bmatrix}^1. \tag{2.16}$$

Then let $s_0 \in \mathbb{C}$ and $k = m_1 + m_2 + \cdots + m_q$ and the matrix $\tilde{V}$ be a matrix (2.16) which is built by SPRIM algorithm so the first $2q$ moments in the expansions of $H$ and the reduced model $\tilde{H}$ around the $s_0$ are identical: $H(s) = \tilde{H}(s) + O((s - s_0)^{2q})$. For the proof see [24].

Also the SPRIM reduced-order model $\tilde{H}$ given by (2.11) (or equivalently, (2.15)) is passive, For the proof see [23].

## 2.2.2 Methods based on Hankel norm approximations and TBR

We consider a brief overview on TBR then we give more details on a recent projection based reduction technique, PMTBR, which uses the advantages of both TBR and projection methods.

[1] After splitting the rows, the columns are not orthogonal anymore hence some reorthogonalization is adopted here.

### Introduction to truncated balanced realization (TBR)

Another group of MOR algorithms suitable for the analysis of LTI systems is based on Hankel norm approximations and TBR [38, 46, 52]. Unlike Krylov-based methods, balanced realization methods have provable error bounds for the reduced order models [46, 49], and guarantee that stability of the original system is preserved in the reduced order model. The main drawback of this class of MOR methods is the high computational cost of extracting the reduced models, associated with expensive ($O(n^3)$) solution of Lyapunov equations. Due to this fast growing complexity, applicability of TBR methods for large-scale systems was limited. In order to overcome this difficulty methods based on approximate gramian computation have been developed [5, 38] and sparse matrix techniques were adopted [8]. The balanced truncation approach, or TBR [46, 49, 58], is centered around the information obtained from the controllability gramian $P$, which can be obtained by solving the Lyapunov equation

$$G \cdot P + P \cdot G^T = -B \cdot B^T.$$

The observability gramian $Q$ can be obtained by solving the dual Lyapunov equation

$$G^T \cdot Q + Q \cdot G = -L^T \cdot L.$$

Under a similarity transformation of the state-space model

$$G \rightarrow T^{-1} \cdot G \cdot T, \quad B \rightarrow T^{-1} \cdot B, \quad L \rightarrow L \cdot T,$$

the input-output properties of the state-space model, such as the transfer function, are invariant. However, the gramians are not invariant as they vary under the transformation

$$P \rightarrow T^{-1} \cdot P \cdot T^{-T}, \quad Q \rightarrow T^T \cdot Q \cdot T.$$

One of the key facts of the TBR procedure is that the eigenvalues of the product of the gramians $P \cdot Q$ do not change. These Hankel singular values contain useful information about the input-output behavior of the system. In particular, the *small* eigenvalues correspond to internal states that have a weak effect on the input-output response of the system and are, therefore, close to non-observable, non-controllable or both. The second key fact is that the gramians are transformed under congruence, and any two symmetric matrices can be simultaneously diagonalized by an appropriate congruence transformation. So it is possible to find a similarity transformation $T$ that leaves the state-space system dynamics unchanged, but transforms the gramians into $\tilde{P}$ and $\tilde{Q}$ equal and diagonal. So in these new coordinates we may apply the partition

$$\tilde{P} = \tilde{Q} = \Sigma = \begin{bmatrix} \Sigma_1 & 0 \\ 0 & \Sigma_2 \end{bmatrix}.$$

$\Sigma_1$ is related to the *strong* state which have relevant effect on the input-output behavior and $\Sigma_2$ is related to the *weak* ones with small effect on the input-output response. The transformed matrices can be partitioned in the same way

$$\tilde{G} = \begin{bmatrix} \tilde{G}_{11} & \tilde{G}_{12} \\ \tilde{G}_{21} & \tilde{G}_{22} \end{bmatrix}, \quad \tilde{B} = \begin{bmatrix} \tilde{B}_1 \\ \tilde{B}_2 \end{bmatrix}, \quad \tilde{L} = \begin{bmatrix} \tilde{L}_1 & \tilde{L}_2 \end{bmatrix}.$$

We can truncate the system by retaining the part that has a *stronger* effect on the input-output behavior, and deleting the part related to *weaker* state. In this way we obtain a reduced system of size $q \ll n$, see (2.7).

Such a model is a reduced order representation of the original system, retaining by construction the most relevant input-output behavior. One of the main advantages of the TBR approach is that the procedure provides *a-posteriori* error bounds on the truncation [46, 49]. The existence of such bounds is quite relevant as it provides a clear way to exchange accuracy for simplicity in the representation. Unfortunately, the balancing procedure is expensive, requiring the solution of the two Lyapunov equations and the eigenvalue decomposition of the transformed gramians. For these reasons, TBR is usually not used for large scale systems.

PMTBR is a projection MOR technique that exploits the direct relation between the multi-point rational projection framework and the TBR. We will give a brief introduction to the method. More details can be found in [48, 58]. We start with a system of the form:

$$\begin{cases} \frac{dx(t)}{dt} &= Ax(t) + Bu(t), \\ y(t) &= Cx(t). \end{cases}$$

For simplicity consider the case $A = A^T$, $C = B^T$ and further assume that $A$ is stable, see Definition 1.2. It is easy to see that in this symmetric case, both observability and controllability gramian are equal. The gramians $X$ can be computed in the time domain via the expression (see Section 1.1.5)

$$X = \int_0^\infty e^{At} B B^T e^{A^T t} dt.$$

Applying Parseval's theorem we obtain an equivalent frequency based expression for the gramian

$$X = \int_{-\infty}^{\infty} (i\omega I - A)^{-1} B B^T (i\omega I - A)^{-*} d\omega,$$

where the superscript $*$ denotes Hermitian transpose and $i = \sqrt{-1}$. Applying a numerical quadrature scheme with nodes $\omega_k$, weights $w_k$ and defining:

$$z_k = (i\ \omega_k\ I - A)^{-1}\ B,$$

an approximation $\hat{X}$ to the gramian $X$ can be computed as:

$$\begin{aligned} X \approx \hat{X} &= \sum_k w_k\ z_k\ z_k^*, \\ &= ZW \cdot (ZW)^*, \\ &= V_z S_z^2 V_z^*. \end{aligned}$$

Here $Z = (z_1, z_2, \ldots)$ and $W = \text{diag}(\sqrt{w_1}, \sqrt{w_2}, \ldots)$; furthermore $S_z$ and $V_z$ are both from the singular value decomposition on $Z \cdot W = V_z \cdot S_z \cdot U_z$. If we use a high order quadrature rule $\hat{X}$ will converge to $X$, which implies that the dominant eigenspace of $\hat{X}$ converges to the dominant eigenspace of $X$. The column vectors in $V_z$ yield the eigenvectors of $\hat{X}$. If an adequate quadrature rule has been chosen, $V_z$ converges to the eigenspace of $X$ and the Hankel Singular Values are obtained directly from the diagonal entries of $S_z$. The dominant eigenvectors of $V_z$ corresponding to the dominant eigenvalues in $S_z$ can be used as a projection matrix via a congruence transformation. The eigenvalues in $S_z$ yield an a-priori error estimation in the way the Hankel Singular Values are used in the TBR procedures for error control.

## 2.3 Homotopized DAEs

For large systems of ordinary differential equations (ODEs), efficient MOR methods already exist in the linear case, see [1]. We want to generalize according techniques to the case of differential algebraic equations (DAEs). On the one hand, a high-index DAE problem can be converted into a lower-index system by analytic differentiations, see [2]. A transformation to index zero yields an equivalent system of ODEs. On the other hand, a regularization is directly feasible in case of semi-explicit systems of DAEs. Thereby, we obtain a singular perturbed problem of ODEs with an artificial parameter. Thus according MOR techniques can be applied to the ODE system. An MOR approach for DAEs is achieved by considering the limit to zero of the artificial parameter.

Systems of DAEs result in the mathematical modeling of a wide variety of problems like electric circuit design, for example. We consider a semi-explicit system

$$\begin{aligned} y'(t) &= f(y(t), z(t)), \qquad & y &: \mathbb{R} \to \mathbb{R}^k \\ 0 &= g(y(t), z(t)), \qquad & z &: \mathbb{R} \to \mathbb{R}^l \end{aligned} \tag{2.17}$$

with differential and perturbation index 1 or 2. For the construction of numerical methods to solve initial value problems of (2.17), the direct as well as the indirect approach can be used. The direct approach applies an *ε-embedding* of the DAEs (2.17), i.e., the system changes into

$$\begin{aligned} y'(t) &= f(y(t), z(t)) \\ \varepsilon z'(t) &= g(y(t), z(t)) \end{aligned} \quad \Leftrightarrow \quad \begin{aligned} y'(t) &= f(y(t), z(t)) \\ z'(t) &= \tfrac{1}{\varepsilon} g(y(t), z(t)) \end{aligned} \tag{2.18}$$

with a real parameter $\varepsilon \neq 0$. Techniques for ODEs can be employed for the singularly perturbed system (2.18). The limit $\varepsilon \to 0$ yields an approach for solving the DAEs (2.17). The applicability and quality of the resulting method still has to be investigated.

Alternatively, the indirect approach is based on the *state space form* of the DAEs (2.17) with differential and perturbation index 1 or 2, for nonlinear cases see [59], i.e.,

$$y'(t) = f(y(t), \Phi(y(t))) \tag{2.19}$$

with $z(t) = \Phi(y(t))$. To evaluate the function $\Phi$, the nonlinear system

$$g(y(t), \Phi(y(t))) = 0 \tag{2.20}$$

is solved for given value $y(t)$. Consequently, the system (2.19) represents ODEs for the differential variables $y$ and ODE methods can be applied. In each evaluation of the right-hand side in (2.19), a nonlinear system (2.20) has to be solved. More details on techniques based on the $\varepsilon$-embedding and the state space form can be found in [35].

Although some MOR methods for DAEs already exist, several techniques are restricted to ODEs or exhibit better properties in the ODE case in comparison to the DAE case. The direct or the indirect approach enables the usage of MOR schemes for ODEs (2.18) or (2.19), where an approximation with respect to the original DAEs (2.17) follows. The aim is to obtain suggestions for MOR schemes via these strategies, where the quality of the resulting approximations still has to be analyzed in each method.

In this work, we focus on the direct approach for semi-explicit system of DAEs, i.e., the $\varepsilon$-embedding (2.18) is considered. MOR methods are applied to the singularly perturbed system (2.18). Two scenarios exist to achieve an approximation of the behavior of the original DAEs (2.17) by MOR. Firstly, an MOR scheme can be applied to the system (2.18) using a constant $\varepsilon \neq 0$, which is chosen sufficiently small (on a case by case basis) such that a good approximation is obtained. Secondly, a parametric MOR method yields a reduced description of the system of ODEs, where the parameter $\varepsilon$ still represents an independent variable. Hence the limit $\varepsilon \to 0$ causes an approach for an approximation of the original DAEs.

We investigate the two approaches with respect to MOR methods based on an approximation of the transfer function, which describes the input-output behavior of the system in frequency domain. We present numerical simulations using two linear semi-explicit systems of DAEs, which follow from mathematical models of electric circuits, see Chapter 4.

## 2.4 Model Order Reduction and $\varepsilon$-Embedding

For the moment we restrict to semi-explicit DAE systems of the type (2.1) and introduce $w(t)$ as an output instead of $y(t)$ with exact the same condition. According to (2.17), the solution $x$ and the matrix $C$ exhibit the partitioning:

$$x = \begin{pmatrix} y \\ z \end{pmatrix}, \qquad C = \begin{pmatrix} I_{k\times k} & 0 \\ 0 & 0_{l\times l} \end{pmatrix}.$$

The order of the system is $n = k+l$, where $k$ and $l$ are the dimensions of the differential part and the algebraic part (constraints), respectively, defined in the semi-explicit system (2.17). A linear system of the form (2.1) is often referred to as the representation of the system in the time domain, or in the state space. As we already have shown in Section 2.1 one can also represent the system in the frequency domain via the Laplace transform (2.4). The corresponding $p \times m$ matrix-valued rational function is:

$$H(s) = L \cdot (G + sC)^{-1} \cdot B = L \cdot \left( G + s \begin{pmatrix} I_{k \times k} & 0 \\ 0 & 0_{l \times l} \end{pmatrix} \right)^{-1} \cdot B$$

provided that $\det(G + sC) \neq 0$. Following the direct approach, the $\varepsilon$-embedding changes the system (2.1) to:

$$\begin{cases} C(\varepsilon)\frac{dx(t)}{dt} & = & -Gx(t) + Bu(t), \qquad x(0) = x_0, \\ w(t) & = & Lx(t), \end{cases} \tag{2.21}$$

where

$$C(\varepsilon) = \begin{pmatrix} I_{k \times k} & 0 \\ 0 & \varepsilon I_{l \times l} \end{pmatrix} \qquad \text{for } \varepsilon \in \mathbb{R}$$

with the same inner state and input/output as before. For $\varepsilon \neq 0$, the matrix $C$ is regular in (2.21) and the transfer function reads:

$$H_\varepsilon(s) = L \cdot (G + s \cdot C(\varepsilon))^{-1} \cdot B$$

provided that $\det(G + sC(\varepsilon)) \neq 0$. For convenience, we introduce the notation

$$M(s, \varepsilon) := sC(\varepsilon) = s \begin{pmatrix} I_{k \times k} & 0 \\ 0 & \varepsilon I_{l \times l} \end{pmatrix}.$$

It holds $M(s, 0) = sC$ with $C$ from (2.1).

Concerning the relation between the original system (2.1) and the regularized system (2.21) with respect to the transfer function, we achieve the following statement. Without loss of generality, the induced matrix norm of the Euclidean vector norm is applied.

**Lemma 2.1.** *Let* $A,\ \tilde{A} \in \mathbb{R}^{n \times n}$, $\det(A) \neq 0$ *and* $\|A - \tilde{A}\|_2 = \|\Delta A\|_2$ *where* $\Delta A$ *is small enough. Then it holds:*

$$\|A^{-1} - \tilde{A}^{-1}\|_2 \leq \frac{\|A^{-1}\|_2^2 \cdot \|\Delta A\|_2}{1 - \|A^{-1}\|_2 \cdot \|\Delta A\|_2}.$$

*Proof.* It holds

$$\|A^{-1} - \tilde{A}^{-1}\|_2 = \max_{\|x\|_2 = 1} \left\| A^{-1}x - \tilde{A}^{-1}x \right\|_2.$$

Suppose $y := A^{-1}x,\ \tilde{y} := \tilde{A}^{-1}x$, then the sensitivity analysis of linear systems yields

$$\frac{\|\Delta y\|_2}{\|y\|_2} \leq \frac{\kappa(A)}{1-\kappa(A)\frac{\|\Delta A\|_2}{\|A\|_2}} \left( \frac{\|\Delta A\|_2}{\|A\|_2} + \underbrace{\frac{\|\Delta x\|_2}{\|x\|_2}}_{=0} \right)$$

where the quantity

$$\kappa(A) \equiv \left\|A^{-1}\right\|_2 \|A\|_2$$

is the relative *condition number*. So by substituting the value of $\kappa(A)$ we have:

$$\|y-\tilde{y}\|_2 \leq \frac{\|A^{-1}\|_2 \cdot \|\Delta A\|_2 \cdot \|A^{-1}\|_2 \, \|x\|_2}{1-\|A^{-1}\|_2 \cdot \|\Delta A\|_2}$$

then

$$\|A^{-1}-\tilde{A}^{-1}\|_2 \leq \frac{\|A^{-1}\|_2^2 \cdot \|\Delta A\|_2}{1-\|A^{-1}\|_2 \cdot \|\Delta A\|_2}.$$

□

We conclude from Lemma 2.1 that

$$\lim_{\Delta A \to 0} \tilde{A}^{-1} = A^{-1}.$$

**Theorem 2.3.** *For fixed* $s \in \mathbb{C}$ *with* $\det(G+M(s,0)) \neq 0$ *and* $\varepsilon \in \mathbb{R}$ *satisfying*

$$|s| \cdot |\varepsilon| \leq \frac{c}{\|(G+M(s,0))^{-1}\|_2} \tag{2.22}$$

*for some* $c \in (0,1)$*, the transfer functions* $H(s)$ *and* $H_\varepsilon(s)$ *of the systems* (2.17) *and* (2.21) *exist and it holds*

$$\|H(s)-H_\varepsilon(s)\|_2 \leq \|L\|_2 \cdot \|B\|_2 \cdot K(s) \cdot |s| \cdot |\varepsilon|$$

*with*

$$K(s) = \frac{1}{1-c} \left\|(G+M(s,0))^{-1}\right\|_2^2.$$

*Proof.* The condition (2.22) guarantees that the matrices $G+M(s,\varepsilon)$ are regular. The definition of the transfer functions implies:

$$\|H(s)-H_\varepsilon(s)\|_2 \leq \|L\|_2 \cdot \left\|(G+M(s,0))^{-1}-(G+M(s,\varepsilon))^{-1}\right\|_2 \cdot \|B\|_2 .$$

Applying the Lemma 2.1, the term at the right-hand side of the expression above becomes:

$$\begin{aligned}\|(G+M(s,0))^{-1}-(G+M(s,\varepsilon))^{-1}\|_2 &\leq \frac{\|(G+M(s,0))^{-1}\|_2^2\cdot\|M(s,0)-M(s,\varepsilon)\|_2}{1-\|(G+M(s,0))^{-1}\|_2\cdot\|M(s,0)-M(s,\varepsilon)\|_2}\\ &\leq \frac{1}{1-c}\left\|(G+M(s,0))^{-1}\right\|_2^2\cdot\|M(s,0)-M(s,\varepsilon)\|_2\\ &\leq K(s)\,\|M(s,0)-M(s,\varepsilon)\|_2\,.\end{aligned}$$

We obtain:

$$\|M(s,0)-M(s,\varepsilon)\|_2 = |s|\cdot\left\|\begin{pmatrix}0 & 0\\ 0 & \varepsilon I_{l\times l}\end{pmatrix}\right\|_2 = |s|\cdot|\varepsilon|\,.$$

Thus the proof is completed. □

It is clear that for inequality (2.22) we have:

$$\begin{aligned} s\neq 0 \ \in\mathbb{C} \ &: \ |\varepsilon|\leq\frac{c}{|s|\cdot\|(G+M(s,0))^{-1}\|_2}\\ s=0 \ \in\mathbb{C} \ &: \ \varepsilon \text{ arbitrary}\end{aligned}$$

We conclude from Theorem 2.3 that

$$\lim_{\varepsilon\to 0} H_\varepsilon(s) = H(s)$$

for each $s\in\mathbb{C}$ with $G+sC$ regular. The relation (2.22) gives the feasible domain of $\varepsilon$

$$\begin{aligned} |s|\leq 1 \ &: \ |\varepsilon|\leq\frac{c}{\|(G+M(s,0))^{-1}\|_2},\\ |s|>1 \ &: \ |\varepsilon|\leq\frac{c}{|s|\cdot\|(G+M(s,0))^{-1}\|_2}.\end{aligned}$$

We also obtain the uniform convergence

$$\|H(s)-H_\varepsilon(s)\|_2\leq\hat{K}\,|\varepsilon| \quad \text{for all} \ \ s\in S$$

in a compact domain $S\subset\mathbb{C}$ and $\varepsilon\leq\delta$ with:

$$\begin{aligned}\delta &= c\cdot\min_{s\in S}\frac{1}{\|(G+M(s,0))^{-1}\|_2} \quad \text{for} \ \ \tilde{S}=\emptyset,\\ \delta &= c\cdot\left[\min_{s\in S}\frac{1}{\|(G+M(s,0))^{-1}\|_2}\right]\cdot\underbrace{\left[\min_{s\in\tilde{S}}\frac{1}{|s|}\right]}_{\leq 1} \quad \text{for} \ \ \tilde{S}\neq\emptyset,\end{aligned}$$

with $\tilde{S}:=\{z\in S:|z|\geq 1\}$. Furthermore, Theorem 2.3 implies the property

$$\lim_{s\to 0} H(s)-H_\varepsilon(s)=0$$

for fixed $\varepsilon$ assuming $\det G \neq 0$. However, we are not interested in the limit case of small variables $s$.

For reducing the DAE system (2.1), we have two ways to handle the artificial parameter $\varepsilon$, which results in two different scenarios. In the first scenario, we fix a small value of the parameter $\varepsilon$. Thus we use one of the standard techniques for the reduction of the corresponding ODE system. Finally, we achieve a reduced ODE (with small $\varepsilon$ inside). The ODE system with small $\varepsilon$ represents a regularized DAE. Any reduction scheme for ODEs is feasible. Figure 2.1 indicates the steps for the first scenario. Recent research shows that the Poor Man's TBR (PMTBR), see [48], can be applied efficiently if the matrix $C$ in (2.1) is regular, which indeed happens in this example. The test case with the simulation result will be presented in Chapter 4.

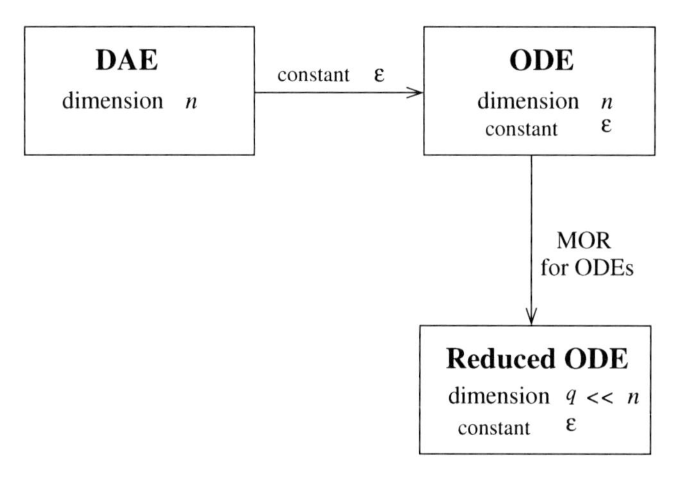


Figure 2.1: The approach of the $\varepsilon$-embedding for MOR in the first scenario.

In the second scenario, the parameter $\varepsilon$ is considered as an independent variable (value not predetermined). We can use the parametric MOR for reducing the corresponding ODE system. The applied parametric MOR is based on [15,20] in this case. The limit $\varepsilon \to 0$ yields the results in an approximation of original DAEs (2.1). The existence of the approximation in this limit still has to be analyzed. Figure 2.2 illustrates the strategy for the second scenario.

Theorem 2.3 provides the theoretical background for the both scenarios. We apply an MOR scheme based on an approximation of the transfer function to the system of ODEs (2.21). Let $\tilde{H}_\varepsilon(s)$ be a corresponding approximation of $H_\varepsilon(s)$.

It follows

$$\|H(s) - \tilde{H}_\varepsilon(s)\|_2 \leq \|H(s) - H_\varepsilon(s)\|_2 + \|H_\varepsilon(s) - \tilde{H}_\varepsilon(s)\|_2 \tag{2.23}$$

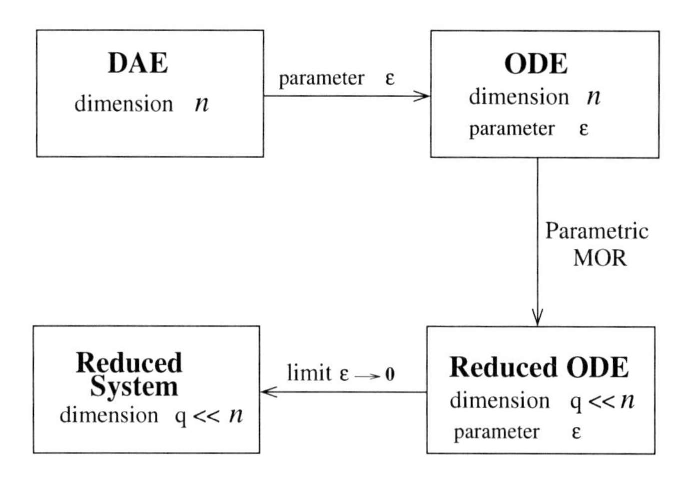


Figure 2.2: The approach of the $\varepsilon$-embedding for MOR in the second scenario.

for each $s \in \mathbb{C}$ with $\det(G+sC) \neq 0$. Due to Theorem 2.3, the first term becomes small for sufficiently small parameter $\varepsilon$. However, $\varepsilon$ should not be chosen smaller than the machine precision on a computer. The second term depends on the applicability of an efficient MOR method to the ODEs (2.21). Thus $\tilde{H}_\varepsilon(s)$ can be seen as an approximation of the transfer function $H(s)$ belonging to the system of DAEs (2.1).

## 2.5 General Linear Systems

The simplest and best understood DAEs are linear equations of the form (2.1). We will focus on this kind of problems to generalize the direct approach ($\varepsilon$-embedding) from Section 2.3. Thus we consider an arbitrary singular matrix $C$ now.

### 2.5.1 Transformation to Kronecker Form

Here investigations are closely related to the theory of matrix pencils, see [7, 63]. This field provides the proposition that the linear DAE (2.1) is uniquely solvable if and only if the matrix pencil $\{C, G\}$ is regular, i.e. the polynomial $\det(\lambda C + G)$ does not vanish identically. We consider constant coefficient matrices $C, G \in \mathbb{R}^{n \times n}$ and the $C^r$-mapping $u : [t_0, t_1] \to \mathbb{R}^m$ represents a time-dependent source term. Due to the regular matrix pencil, the matrices $G$ and $C$ can be transformed simultaneously to the *Kronecker canonical form*, see [35],

$$PCQ = \begin{pmatrix} I_{n-m} & 0 \\ 0 & N \end{pmatrix}, \quad PGQ = \begin{pmatrix} M & 0 \\ 0 & I_m \end{pmatrix} \tag{2.24}$$

with the regular matrices $P, Q \in \mathbb{R}^{n\times n}$. It holds $M \in \mathbb{R}^{(n-m)\times(n-m)}$ and $N \in \mathbb{R}^{m\times m}$ is a nilpotent matrix with the nilpotency index $\nu$, i.e., $N^\nu = 0$ but $N^{\nu-1} \neq 0$. If $C$ is regular, then we have $\nu = 0$. Since we suppose that the matrix $C$ in (2.1) is singular, the matrix $N$ in (2.24) exhibits the following structure

$$N = \begin{pmatrix} 0 & * & \dots & * \\ \vdots & \ddots & \ddots & \vdots \\ \vdots & & \ddots & * \\ 0 & \dots & \dots & 0 \end{pmatrix}. \tag{2.25}$$

More precisely, just one upper diagonal may be occupied. The special structure of $N$ in (2.25) allows us to add the variable $\varepsilon$ on the diagonal of the matrix $N$, which corresponds to a regularization of the product $PCQ$. Without this transformation the matrix $C$ does not have any special pattern in general to add this variable $\varepsilon$ to avoid the singularity. The transformation of the pencil $\{C, G\}$ to its Kronecker canonical form corresponds to a decoupling of the DAE (2.1) into

$$\dot{y} + My = \eta(t), \tag{2.26a}$$

$$N\dot{z} + z = \delta(t) \tag{2.26b}$$

with

$$x = Q \cdot \begin{pmatrix} y \\ z \end{pmatrix}, \quad P \cdot B \cdot u(t) = \begin{pmatrix} \eta(s) \\ \delta(t) \end{pmatrix}. \tag{2.26c}$$

Now (2.26a) is already an explicit ODE for $y$. We can obtain an ODE for $z$ from (2.26b) as follows, see [35]. One differentiation and multiplication by $N$ yields,

$$N^2\ddot{z} + N\dot{z} = N\dot{\delta}(t) \overset{(2.26b)}{\Rightarrow} z = \delta(t) - N\dot{\delta}(t) + N^2\ddot{z}.$$

Proceeding in this way, i.e. by successive differentiation and multiplication by $N$, we arrive after $\nu - 1$ steps (because $N^\nu = 0$) at

$$z = \delta(t) - N\dot{\delta}(t) + N^2\ddot{\delta}(t) - \dots + (-1)^\nu N^{\nu-1}\delta^{(\nu)}. \tag{2.27}$$

One differentiation of (2.27) results in an ODE for $z$. Hence the differential index is $\nu$, presumed that $\delta$ is sufficiently often differentiable. We define $v := Q^{-1}x$. Applying the Kronecker canonical form (2.24) changes the linear DAEs (2.1) into the transformed linear system

$$\begin{cases} PCQ\frac{\mathrm{d}v(t)}{\mathrm{d}t} &= -PGQv(t) + PBu(t) \\ w(t) &= LQv(t). \end{cases} \tag{2.28}$$

According to (2.24), we define $\hat{C} := PCQ$. Remark that both $C$ and $N$ are singular matrices. The representation of the system in the frequency domain via the Laplace transform (2.4) results to

$$H(s) = LQ \cdot (PGQ + s\underbrace{PCQ}_{\hat{C}})^{-1} \cdot PB. \tag{2.29}$$

Following the direct approach, the $\varepsilon$-embedding changes the system (2.28) into

$$\begin{cases} \hat{C}_\varepsilon \frac{\mathrm{d}v(t)}{\mathrm{d}t} &= -PGQv(t) + PBu(t) \\ w(t) &= LQv(t) \end{cases} \tag{2.30}$$

with

$$\hat{C}_\varepsilon := \begin{pmatrix} I_{n-m} & 0 \\ 0 & \varepsilon I_m + N \end{pmatrix} \quad \text{for } \varepsilon \neq 0 \tag{2.31}$$

and the same inner state and input/output as before. For $\varepsilon \neq 0$, the matrix $\hat{C}_\varepsilon$ is regular in (2.31) and the transfer function of (2.30) reads

$$H_\varepsilon(s) = LQ \cdot (PGQ + s\hat{C}_\varepsilon)^{-1} \cdot PB. \tag{2.32}$$

Concerning the relation between the original system (2.28) and the regularized system (2.30) with respect to the transfer function, we achieve the following statement. Without loss of generality, the induced matrix norm of the Euclidean vector norm is applied again.

**Theorem 2.4.** *For fixed $s \in \mathbb{C}$ with $\det(PGQ + s\hat{C}) \neq 0$ and $\varepsilon \in \mathbb{R}$ satisfying*

$$|s| \cdot |\varepsilon| \leq \frac{c}{\|(PGQ + s\hat{C})^{-1}\|_2} \tag{2.33}$$

*for some $c \in (0,1)$, the transfer functions $H(s)$ from (2.29) and $H_\varepsilon(s)$ from (2.32) exist and it holds*

$$\|H(s) - H_\varepsilon(s)\|_2 \leq \|L\|_2 \cdot \|B\|_2 \cdot \|P\|_2 \cdot \|Q\|_2 \cdot K(s) \cdot |s| \cdot |\varepsilon|$$

*with*

$$K(s) = \frac{1}{1-c} \left\| (PGQ + s\hat{C})^{-1} \right\|_2^2 .$$

*Proof.* The condition (2.33) guarantees that the matrices $PGQ + s\hat{C}_\varepsilon$ are regular. The definition of the transfer functions implies

$$\begin{aligned} &\|H(s) - H_\varepsilon(s)\|_2 \\ &\leq \|L\|_2 \cdot \|Q\|_2 \cdot \left\| (PGQ + s\hat{C})^{-1} - (PGQ + s\hat{C}_\varepsilon)^{-1} \right\|_2 \cdot \|P\|_2 \cdot \|B\|_2 . \end{aligned}$$

Applying Lemma 2.1, the term in the above right-hand side satisfies the estimate

$$\begin{aligned} &\left\|(PGQ + s\hat{C})^{-1} - (PGQ + s\hat{C}_\varepsilon)^{-1}\right\|_2 \\ &\leq \frac{1}{1-c}\left\|(PGQ + s\hat{C})^{-1}\right\|_2^2 \cdot \left\|(PGQ + s\hat{C}) - (PGQ + s\hat{C}_\varepsilon)\right\|_2 . \end{aligned}$$

Using basic calculations, it follows

$$\left\|(PGQ + s\hat{C}) - (PGQ + s\hat{C}_\varepsilon)\right\|_2 = |s| \cdot \left\|\hat{C} - \hat{C}_\varepsilon\right\|_2 = |s| \cdot |\varepsilon|.$$

Thus the proof is completed. □

We conclude from Theorem 2.4 that

$$\lim_{\varepsilon \to 0} H_\varepsilon(s) = H(s)$$

for each $s \in \mathbb{C}$ with $G + sC$ regular. Uniform convergence is given in a compact set $s \in S \subset \mathbb{C}$ again. Although the above theorem guarantees the convergence of the transfer function of the regularized system to the transfer function of the original system in the limit case, we encounter two drawbacks.

The first one is related to the numerical calculation of the Kronecker canonical form. The numerical computation of the Kronecker canonical form might be unstable, see [17], due to a possible ill-conditioning of the matrices $P$ and $Q$. The second issue is that the upper bound in Theorem 2.4 includes the norms of $P$ and $Q$ now, since we have applied a transformation before the regularization. If these norms are large, we obtain a pessimistic estimate. The numerical difficulties lead us to seek an alternative just for the numerical calculation although from the theoretical point of view the above approach is feasible.

### 2.5.2 Transformation via Singular Value Decomposition

In the following we will introduce an alternative to the Kronecker canonical form which has no side effect for the numerical implementation. We apply the *singular value decomposition* (SVD), see [62], to the matrix $C$ in the system (2.1). For an arbitrary matrix $M \in \mathbb{R}^{m \times n}$, it exists a factorization of the form

$$UMV^\top = \Sigma, \tag{2.34}$$

where $U \in \mathbb{R}^{m \times m}$ and $V \in \mathbb{R}^{n \times n}$ are orthogonal matrices. The matrix $\Sigma \in \mathbb{R}^{m \times n}$ is diagonal with nonnegative real entries, which are the singular values. The factorization (2.34) is called a singular value decomposition of $M$. Applying the SVD form (2.34) transforms the linear system of DAEs (2.1) into

$$\begin{cases} UCV^\top V \frac{\mathrm{d}x(t)}{\mathrm{d}t} &= -UGV^\top Vx(t) + UBu(t) \\ w(t) &= LV^\top Vx(t). \end{cases}$$

We define $z := Vx$. It follows

$$\left\{ \begin{array}{rcl} \begin{pmatrix} \tilde{\Sigma} & 0 \\ 0 & 0 \end{pmatrix} \frac{\mathrm{d}z(t)}{\mathrm{d}t} & = & -UGV^\top z(t) + UBu(t) \\ w(t) & = & LV^\top z(t), \end{array} \right. \tag{2.35}$$

where the diagonal matrix $\tilde{\Sigma} \in \mathbb{R}^{r \times r}$ $(r < n)$ contains the positive singular values. Thus $\tilde{\Sigma}$ is regular. The $\varepsilon$-embedding changes the system into

$$\left\{ \begin{array}{rcl} \underbrace{\begin{pmatrix} \tilde{\Sigma} & 0 \\ 0 & \varepsilon I_{n-r} \end{pmatrix}}_{C_\varepsilon} \frac{\mathrm{d}z(t)}{\mathrm{d}t} & = & -UGV^\top z(t) + UBu(t) \\ w(t) & = & LV^\top z(t). \end{array} \right. \tag{2.36}$$

The introduced matrix $C_\varepsilon$ is regular. We obtain the same result as for the transformation to Kronecker canonical form.

**Theorem 2.5.** *For fixed $s \in \mathbb{C}$ with $\det(UGV^\top + sUCV^\top) \neq 0$ and $\varepsilon \in \mathbb{R}$ satisfying*

$$|s| \cdot |\varepsilon| \leq \frac{c}{\|(UGV^\top + sUCV^\top)^{-1}\|_2}$$

*for some $c \in (0,1)$, the transfer functions $H(s)$ and $H_\varepsilon(s)$ of the systems* (2.35) *and* (2.36) *exist and it holds*

$$\|H(s) - H_\varepsilon(s)\|_2 \leq \|L\|_2 \cdot \|B\|_2 \cdot K(s) \cdot |s| \cdot |\varepsilon|$$

*with*

$$K(s) = \frac{1}{1-c} \left\| \left( UGV^\top + sUCV^\top \right)^{-1} \right\|_2^2 .$$

The steps of the proof for the previous theorem are like in the proof of Theorem 2.3, see [44], since the matrices after applying the SVD exhibit a semi-explicit structure. Remark that the transfer function of (2.35) coincides with the function of (2.1). The two drawbacks of the transformation to Kronecker canonical form are omitted by the SVD. Firstly, stable numerical methods exist to compute the SVD efficiently. Secondly, we obtain the same upper bound as for the semi-explicit systems. The reason is that the orthogonal matrices feature the optimal property

$$\|U\|_2 = \left\|U^\top\right\|_2 = \|V\|_2 = \left\|V^\top\right\|_2 = 1.$$

Hence the SVD can be used as a brilliant alternative for decoupling the dynamical system. Afterwards, we can apply the $\varepsilon$-embedding.

## 2.6 Parametric Model Order Reduction

Standard MOR-techniques attempt to create order reduced models of large scale systems, arising e.g, in interconnect modeling, that show similar input-output behavior over a wide range of input-frequencies as the full system. However, capturing dependency of the system's behavior w.r.t. variations of other factors is of crucial importance as well. Factors that need to be considered may be design parameter of geometry and operating temperature as well as influence of variations within the design process on the network's performance. Subsuming, say $k$ parameters $\lambda_1, \ldots, \lambda_k$, a linear dynamical system, which is linearly depending on the above parameters can be described by

$$\begin{cases} C(\lambda)\dot{x} + G(\lambda)x &= Bu \\ y &= L^T x, \end{cases}$$

where $\lambda = (\lambda_1, \ldots, \lambda_k)$ and initial conditions $x(\lambda, 0) = x_0(\lambda)$. The matrices $C(\lambda)$ and $G(\lambda)$ are of the form

$$\begin{aligned} C(\lambda) &= C_0 + \lambda_1 C_1 + \cdots + \lambda_k C_k \\ G(\lambda) &= G_0 + \lambda_1 G_1 + \cdots + \lambda_k G_k, \end{aligned} \tag{2.37}$$

where $C_i, G_i \in \mathbb{R}^{n \times n}$. Note that the form (2.37) might also arise from linearization of nonlinear depending system matrices $G(\lambda), C(\lambda)$ around $\lambda = (0, \ldots, 0)$. The one-parameter-case, i.e. having one geometrical parameter plus the frequency $s$ marks the transfer to parameterized model order reduction. In this case, the time domain problem is of the from

$$\begin{cases} (C_0 + \lambda C_1)\dot{x} + (G_0 + \lambda G_1)x &= Bu \\ y &= L^T x \end{cases} \tag{2.38}$$

and after Laplace-transformation one obtains the frequency-domain problem

$$\begin{cases} s(C_0 + \lambda C_1)X(s) + (G_0 + \lambda G_1)X(s) &= BU(s) \\ Y(s) &= L^T X(s) \end{cases} \tag{2.39}$$

resulting in the transfer function

$$H(s, \lambda) = L^T (G_0 + \lambda G_1 + s(C_0 + \lambda C_1))^{-1} B.$$

Projection based MOR techniques now search for a projection matrix $V$ (projecting onto some reduced subspace V), like in the non-parameterized case, such that the system

$$\begin{cases} (\hat{C}_0 + \lambda \hat{C}_1)\dot{z} + (\hat{G}_0 + \lambda \hat{G}_1)z &= \hat{B}u(s) \\ \hat{y} &= \hat{L}^T x \end{cases}$$

of reduced dimension $q < n$ is corresponding in some sense to the full system (2.38). Where $\hat{C}_0, \hat{C}_1, \hat{G}_0, \hat{G}_1$ arise in the standard Galerkin-projection manner

$$\hat{C}_0 = V^T C_0 V, \; \hat{C}_1 = V^T C_1 V, \ldots.$$

In Krylov-based techniques $V$ is constructed such that the moments, i.e. coefficients of the series expansions of $H(s,\lambda)$ and the reduced system's transfer function

$$\hat{H}(s,\lambda) = \hat{L}^T(\hat{G}_0 + \lambda\hat{G}_1 + s(\hat{C}_0 + \lambda\hat{C}_1))^{-1}\hat{B} \tag{2.40}$$

are matched. A series expansion of $H(s,\lambda)$ around $(s,\lambda) = (0,0)$ is given by

$$H(s,\lambda) = \sum_{j=0}^{\infty}\left(\sum_{i=0}^{\infty} m_i^j s^i\right)\lambda^j$$

with the *multi-parameter moments* $m_i^j = L^T q_i^j$ where

$$\begin{cases} q_i^j &= 0 & \text{if} \quad i<0 \quad \text{or} \quad j<0 \\ q_0^0 &= G_0^{-1}b & \text{if} \quad i=j=0 \\ q_i^j &= -G_0^{-1}(C_0 q_{i-1}^j + G_1 q_i^{j-1} + C_1 q_{i-1}^{j-1}) & \text{otherwise.} \end{cases}$$

for more details, see [41]. Various strategies to achieve moment matching can be found in literature. In the following we give just a brief overview and references for further studies:

1. In view of (2.40) one can introduce an auxiliary parameter $\mu = s\lambda$ and rewrite the transfer function

   $$H(s,\lambda,\mu) = L^T(G_0 + sC_0 + \lambda G_1 + \mu C_1)^{-1}B$$

   and match moments of the corresponding Taylor series expansion

   $$H(s,\lambda,\mu) = \sum_{k=0}^{\infty}\sum_{j=0}^{k}\sum_{i=0}^{k-j} m_{i,j,k}\lambda^{k-(i+j)}s^i\mu^j.$$

   To match the moments $m_i^j$ for $i = 0,1,\ldots,p$ and $j = 0,1,\ldots,q$ the subspace $\nu$ spanned by $V$ has to be $O((p+q)^3)$, see [15, 19].

2. In [39] the approach CORE constructs in a first step a linear system to explicitly match the geometric parameter $\lambda$ to order $q$. Then, with the system matrices of this auxiliary system the Arnoldi algorithm is run to implicitly match $p$ moments w.r.t the frequency parameter $s$. The necessary dimension $p+1$ of the final reduced system is accompanied by a numerical instability, missing structure preservation and lack of passivity preservation.

3. PIMTAP, proposed in [41] suggests to construct the subspace $\mathcal{V}$ on which the system is projected from:

   $$\mathcal{V} = \text{span}\begin{cases} q_0^0 & q_1^0 & \cdots & q_p^0 \\ \vdots & & & \\ q_0^q & q_1^q & \cdots & q_p^q. \end{cases} \tag{2.41}$$

Here an orthogonal basis (which forms $V$) is constructed column wise from (2.41) using a so-called 2D Arnoldi process. In [41] it is furthermore stated that this approach constructs $V$ in a numerically stable way, the structure (2.38) as well as stability and passivity is also preserved.

4. The method suggested in [33] starts somehow from a different point of view. Revisiting (2.39) we recognize that, because the system matrices depend on the geometric parameter $\lambda$, also the state $X$ depends on $\lambda$, i.e. $X = X(s, \lambda)$. From this starting point, moments, i.e. derivatives, w.r.t. one parameter in $\{s, \lambda\}$ are computed by fixing the other one and evaluating coefficients.

5. Yet another approach is given in [19]. Starting from a system with two parameters i.e. one geometric parameter plus frequency [1]

$$\begin{cases} (C_0 + sC_1 + \lambda C_2)X(s, \lambda) &= BU(s, \lambda) \\ Y(s, \lambda) &= L^T X(s, \lambda) \end{cases}$$

where $C_0$, $C_1$ and $C_2 \in \mathbb{R}^{n \times n}$ are system matrices, the transfer function

$$H(s, \lambda) = L^T (C_0 + sC_1 + \lambda C_2)^{-1} B$$

is first expanded into a Taylor series w.r.t. the parameter $s$ keeping $\lambda$ constant. Without giving details, which can be found in [19], it is clear that the moments, i.e. Taylor series coefficients, of this expansion depend on the free parameter $\lambda$. Hence, in a second step the moments are expanded w.r.t. the second parameter $\lambda$. It turns out, that the final coefficients of this series expansion are independent of the parameters and are, therefore, used to define the projection matrix $V$.

## Parameterized interconnect macromodeling via a two-directional Arnoldi process

The extension of Krylov model order reduction (KMOR) with one parameter to multiple parameters is difficult. It is much more complicated both in complexity and numerical stability. Pioneering work includes [19, 21, 33, 39, 75]. Parameterized interconnect macro-modeling via a two-directional Arnoldi process (PIMTAB) [40–42] provides a flexible, systematic and relatively numerically stable way for reducing linear systems with multiple parameters. The linear system that PIMTAB solves in the two-parameter case is of the form (2.39). The Taylor expansion of $X$ becomes:

$$X = \sum_{i=0}^{\infty} \sum_{j=0}^{\infty} q_i^j s^i \lambda^j, \ (q_i^j \in \mathbb{C}^N),$$

where $q_i^j$ is called the $(i, j)$-th 2-parameter moment of $X$.

[1] comparing to [39] the mixed term (with $s\lambda$) is missing.

By substituting this Taylor series in (2.39) and some basic calculations we obtain the recursive relationship of the moments:

PIMTAP projects (2.39) to a subspace $\mathcal{V}$ spanned by a selection of the moments. PIMTAP has two implementations: PIMTAP via Arnoldi process and PIMTAP via Two-directional Arnoldi Process (TAP). The subspace that PIMTAP via Arnoldi Process uses for projection is $\mathcal{V}(p,q) = \text{span}\{r_i^j : i = 0, \ldots, p-1, j = 0, \ldots, q-1\}$, and if we define $q_{[i]}^{[j]}$, $G_{[j]}$ and $C_{[j]}$ as

$$q_{[i]}^{[j]} = \begin{bmatrix} q_{i-1}^0 \\ q_{i-1}^1 \\ \vdots \\ q_{i-1}^{j-1} \end{bmatrix}, G_{[j]} = \underbrace{\begin{bmatrix} G_0 & & & & \\ G_1 & G_0 & & & \\ & G_1 & G_0 & & \\ & & \ddots & \ddots & \\ & & & G_1 & G_0 \end{bmatrix}}_{j \text{ blocks}}, C_{[j]} = \underbrace{\begin{bmatrix} C_0 & & & & \\ C_1 & C_0 & & & \\ & C_1 & C_0 & & \\ & & \ddots & \ddots & \\ & & & C_1 & C_0 \end{bmatrix}}_{j \text{ blocks}},$$

the recursive relationship could be rearranged as

$$q_{[i]}^{[j]} = -G_{[j]}^{-1} C_{[j]} q_{[i-1]}^{[j]}, \quad \text{for all } i > 1. \tag{2.42}$$

From the relationship (2.42), we can see that $q_{[1]}^{[j]}, q_{[2]}^{[j]}, \cdots, q_{[k]}^{[j]}$ span a Krylov subspace $\mathcal{K}_k\left\{-G_{[j]}^{-1} C_{[j]}, q_{[i]}^{[j]}\right\}$, which can be generated with the standard Arnoldi process. Splitting each Arnoldi vector to $j$ vectors in $\mathbb{C}$ and then orthogonalizing and normalizing all these vectors, we get the projection matrix $V \in \mathbb{C}^{n \times kj}$ whose column vectors span $\mathcal{V}(k,j)$.

A restriction of PIMTAP via Arnoldi process is that the moment matching pattern must be a rectangle as is shown in Figure 2.3(a), in which a solid circle in $(i,j)$ means $q_i^j$ is to be matched. In some applications, the high order cross-term moments are not so important. An extreme example is the method proposed in [33] that matches none of the cross-term moments but is still accurate enough for its test cases. Therefore, PIMTAP via TAP is proposed in [40–42] to generate more flexible moment matching patterns like the example shown in Figure 2.3(b). The only restriction of moment matching pattern in PIMTAP via TAP is that if $r_m^n$ is matched, any $q_i^j$ $(0 \leq i < m, 0 \leq j < n)$ must also be matched. Thus, the moment matching pattern could be expressed by vector $p \in \mathbb{N}^k$ satisfying $0 < p(i) \leq p(j)$ when $i > j$, meaning that $q_i^j$ is matched if and only if $j \leq k$ and $i \leq p(j)$. For example, the moment matching vector of the example in Figure 2.3(b), is $(10, 7, 4, 2)$.

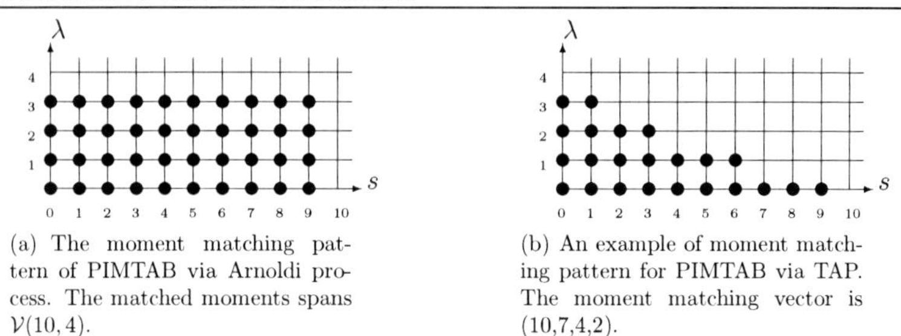


(a) The moment matching pattern of PIMTAB via Arnoldi process. The matched moments spans $\mathcal{V}(10,4)$.

(b) An example of moment matching pattern for PIMTAB via TAP. The moment matching vector is (10,7,4,2).

Figure 2.3: Illustrations of the moment matching patterns of PIMTAP.

The idea of PIMTAP via TAP to generate the moment matching pattern in Figure 2.3(b) is to build four Krylov subspace: $\mathcal{K}^{(r)}(10,1), \mathcal{K}^{(r)}(7,2), \mathcal{K}^{(r)}(4,3), \mathcal{K}^{(r)}(2,4)$[1], split each the Arnoldi vector to vector(s) in $\mathbb{C}^n$ and then orthogonalized them to get the projection matrix $V \in \mathbb{C}^{n\times 23}$. A shortcoming of this method is that some moments are computed repeatedly, and TAP is an improvement of this method by recycling the moments that are already computed. For more details about TAP, please consult [40].

Use the PIMTAP vectors as the column vectors of $V \in \mathbb{C}^{n\times k}$ and let $W \in \mathbb{C}^{n\times k}$ be an arbitrary matrix satisfying that $W^* G_0 V$ is nonsingular, the one-sided reduced order model can be obtained by approximating $x$ by $Vz\,(z \in \mathbb{C}^k)$ and then left multiplying the first equation of (2.39) with $W^*$:

$$\begin{cases} (\hat{G}_0 + \lambda\hat{G}_1 + s(\hat{C}_0 + \lambda\hat{C}_1))Z(s,\lambda) &= \hat{B}U(s), \\ Y(s,\lambda) &= \hat{L}^* Z(s,\lambda), \end{cases}$$

where $\hat{G}_0 = W^*G_0V$, $\hat{G}_1 = W^*G_1V$, $\hat{C}_0 = W^*C_0V$, $\hat{C}_1 = W^*C_1V$, $\hat{B} = W^*B$ and $\hat{L} = V^*L$. After the reduction, the moments specified by the moment matching pattern are matched.

[1] The first numbers in parentheses referred to moment-index in x-direction ($s$) and the second numbers in parentheses referred to moment-index in y-direction ($\lambda$) as an example $(10,1)$ means 10 moments in $s$ direction and just 1 moment in $\lambda$ direction. Also $r$ refers to right Krylov subspace.

# Chapter 3

# Nonlinear model order reduction

In this chapter we review the status of existing techniques for nonlinear model order reduction by investigating how well these techniques perform for circuit simulation. The two best-known methods for reduction of nonlinear systems are proper orthogonal decomposition (POD) [6, 10, 37], also known as Karhunen-Loève expansion [43], and trajectory piecewise-linear techniques (TPWL) [54, 55], which are discussed in Section 3.2 and Section 3.3, respectively. We address several questions that are (closely) related to both the theory and application of nonlinear model order reduction techniques.

## 3.1 Nonlinear versus Linear Model Order Reduction

Electrical circuits may contain nonlinear elements, e. g. nonlinear resistors to model diodes, nonlinear capacitors modeling varactors or combinations of these, corresponding to the behavior of transistors. Hence, the dynamics of electrical circuits can generally be described by a nonlinear, first order, differential-algebraic equation (DAE) of the form:

$$\frac{d}{dt}q(x(t)) + j(x(t)) + Bu(t) = 0; \quad y(t) = L^T x(t), \tag{3.1}$$

where, as in the linear case, $x(t) \in \mathbb{R}^n$ represents the unknown vector of circuit variables at time $t \in \mathbb{R}$ and $B \in \mathbb{R}^{n\times m}$ distributes the input excitation $u : \mathbb{R} \to \mathbb{R}^m$ and $L \in \mathbb{R}^{n\times q}$ maps the state $x$ to the system response $y(t) \in \mathbb{R}^q$. The contribution of nonlinear reactive and nonreactive elements is described by $q, j : \mathbb{R}^n \to \mathbb{R}^n$, respectively.

We interpret (3.1) as a description of a sub-circuit. Thus the input $u$ and the output $y$ are terminal voltages and terminal currents that are injected and extracted linearly. Model order reduction aims at replacing the original model (3.1) by a system

$$\frac{d}{dt}[\tilde{q}(z(t))] + \tilde{j}(z(t)) + \tilde{B}u(t) = 0; \quad \tilde{y}(t) = \tilde{L}^T z(t), \tag{3.2}$$

with $z(t) \in \mathbb{R}^r$; $\tilde{q}, \tilde{j} : \mathbb{R}^r \to \mathbb{R}^r$ and $\tilde{B} \in \mathbb{R}^{r \times m}$ and $\tilde{L} \in \mathbb{R}^{r \times q}$, where we can compute a system response $\tilde{y}(t) \in \mathbb{R}^q$ that is sufficiently close to $y(t)$ given the same input signal $u(t)$, but in much less time.

In the previous chapters, we have described approaches to MOR for linear problems. When transferring these techniques to nonlinear networks, fundamental differences emerge. Especially one realizes that only reduction of the dimension of the problem is not enough to reduce the time needed for solving a system numerically.

To understand this, we recall the basic principle of linear MOR, see Section 2.2, i. e. MOR for linear problems of the form

$$C\frac{d}{dt}x(t) + Gx(t) + Bu(t) = 0; \quad y(t) = L^T x(t) \tag{3.3}$$

with $C, G \in \mathbb{R}^{n \times n}$.

The basic idea of projection based linear MOR is to approximate the state $x(t) \in \mathbb{R}^n$ in a lower dimensional space of dimension $r \ll n$, spanned by basis vectors which we collect in $V = (v_1, \ldots, v_r) \in \mathbb{R}^{n \times r}$:

$$x(t) \approx Vz(t), \quad \text{with} \quad z(t) \in \mathbb{R}^r.$$

The reduced state $z$ is defined by a reduced dynamical system that arises from projecting (3.3) on a test space spanned by the columns of some matrix $W$. There, $W$ and $V$ are chosen biorthonormal, i.e., $W^T V = I_{r \times r}$[1]. The Galerkin projection[2] yields

$$\tilde{C}\frac{d}{dt}z(t) + \tilde{G}z(t) + \tilde{B}u(t) = 0; \quad y(t) = \tilde{L}^T z(t)$$

with $\tilde{C} = W^T CV$, $\tilde{G} = W^T GV \in \mathbb{R}^{r \times r}$ and $\tilde{B} = W^T B \in \mathbb{R}^{r \times m}$, $\tilde{L} = V^T L \in \mathbb{R}^{r \times p}$. The system matrices $\tilde{C}, \tilde{G}, \tilde{B}, \tilde{L}$ of this reduced substitute model are of smaller dimension and constant, i.e., need to be computed only once. We need to guarantee that the reduced pencil is still regular, see [65, 66].

Applying the same technique, e. g. with $V$ and $W$ constructed via POD, directly to the nonlinear system (3.1) means obtaining the reduced formulation (3.2) by defining

$$\tilde{q}(z) = W^T q(Vz) \quad \text{and} \quad \tilde{j}(z) = W^T j(Vz). \tag{3.4}$$

Clearly, $\tilde{q}$ and $\tilde{j}$ map from $\mathbb{R}^r$ into $\mathbb{R}^r$. The original nonlinear d-dimensional DAE model is reduced to a nonlinear r-dimensional DAE reduced order model by means of the Galerkin projection. Unfortunately, the resulting reduced order model for $z \in R^r$ is not always solvable for any arbitrary truncation degree $r$, see [3, 70].

[1] This is a very practical case for an explicit ODE when $C = I$.
[2] Most frequently $V$ is constructed to be orthogonal, such that $W = V$ can be chosen.

Each numerical integration scheme applied to (3.2) calls for evaluations of the functions $\tilde{q}(z_l), \tilde{j}(z_l)$ as well as the system's Jacobian matrix

$$\tilde{J} = \left(\alpha\frac{\partial \tilde{q}}{\partial z} + \frac{\partial \tilde{j}}{\partial z}\right)\Big|_{z=z_l} = W^T\left(\alpha\frac{\partial q}{\partial x} + \frac{\partial j}{\partial x}\right)\Big|_{x=Vz_l} V,$$

where $\alpha \in \mathbb{R}$ is some integration coefficient, at some intermediate points $z_l$.

In the nonlinear case, both the evaluation of the functions and the Jacobians necessitate the back projection, i. e. prolongation of the argument $z_l$ to its counterpart $Vz_l$ followed by the evaluation of the functions $q$ and $j$ and the projection to the reduced space with $W$ and $V$.

Consequently, no reduction will be obtained with respect to computation time unless additional measures are taken or other strategies are pursued.

## 3.2 Proper orthogonal decomposition and adaptations

POD [37] is a method, which extracts principal information from a set of multidimensional data in order to approximate the information contained in the set on a set of lower dimension. Hence, POD suggests itself to be used in MOR to construct the subspace for the projection of the system. Recently also adaptations to MOR with POD where published that overcome the problem with Garlerkin-projected nonlinear problems, see Section 3.1. In the following we give a brief introduction to POD and we refer to [50] for further reading. Furthermore, we sketch some of the adaption, where we give references in the corresponding paragraphs.

### 3.2.1 Basis of POD

In MOR for dynamical systems, the data set, used by POD, is collected during a benchmark simulation. That means, given a typical input signal $u(t)$, the full system (3.1) is solved on some time interval $[t_0, t_{\text{end}}]$. During this simulation, say $K$, snapshots are taken and gathered in a snapshot matrix

$$X = (x_1, \ldots, x_K) \in \mathbb{R}^{n\times K}.$$

The columns of the matrix $X$ span a subspace of dimension $k \leq K$, for which POD creates an optimal orthonormal basis $\{v_1, \ldots, v_k\}$. An orthonormal basis is considered optimal, when the error

$$e = (\|x_1 - \hat{x}_1\|_2^2, \ldots, \|x_k - \hat{x}_k\|_2^2)^T \in \mathbb{R}^k$$

is minimized with respect to the averaging operator

$$\langle e \rangle = \frac{1}{K} \sum_{i=1}^{K} e_i.$$

Here $x_i$ and $\hat{x}_i$ are defined as:

$$\begin{aligned} x_i &= \alpha_{i1} v_1 + \alpha_{i2} v_2 + \cdots + \alpha_{ik} v_k \\ &= \sum_{j=1}^{k} \alpha_{ij} v_j \qquad \text{for} \quad i = 1, \ldots, k, \\ \hat{x}_i &= \sum_{j=1}^{r} \alpha_{ij} v_j \approx x_i \qquad \text{for} \quad i = 1, \ldots, k \\ & \qquad \qquad \qquad \qquad \text{and} \quad r < k. \end{aligned}$$

This is a least squares problem which can be solved by the singular value decomposition (SVD) of the snapshot matrix

$$X = UST \quad \text{with} \quad U \in \mathbb{R}^{n \times n}, \quad T \in \mathbb{R}^{K \times K} \text{ and } S = \left( \begin{matrix} \sigma_1 & & \\ & \ddots & \\ & & \sigma_n \end{matrix} \,\middle|\, 0_{n \times (K-n)} \right),$$

where we assume $K > n$. $U$ and $T$ are orthogonal matrices and the singular values satisfy without loss of generality $\sigma_1 \geq \cdots \geq \sigma_n \geq 0$. The level of truncation is chosen as the smallest $r$ such that for some threshold $d$ (typically $d = 99$) we have

$$\frac{\sum_{i=1}^{r} \sigma_i^2}{\sum_{i=1}^{n} \sigma_i^2} \geq \frac{d}{100}.$$

Now, in MOR with POD, $V$ and $W$ in (3.4) are chosen as $V = W = (v_1, \ldots, v_r)$.

### 3.2.2 Adaption of POD for nonlinear problems

As we have seen, POD does not need any information about the particular structure of the problem, i.e. it does not ask if the problem is nonlinear or not. As it is based on the snapshots only, it may be applied straightforward to nonlinear problems. However, as mentioned before, the cost of evaluating the element functions $q, j$ is not reduced. Adaptations have been made, as we will describe in the following. Common to all the approaches is, to replace the element functions $q, j : \mathbb{R}^n \to \mathbb{R}^n$, i.e., actually a collection of $n$ functions each, by $\widetilde{q}(\cdot), \widetilde{j}(\cdot) \in \mathbb{R}^r$, i.e. a collection of $r < n$ functions. In the adaption presented, $\widetilde{q}$ and $\widetilde{j}$ are a subset of $q$ and $j$, respectively. They differ in the way, the selection is chosen.

### Missing point estimation

Missing point estimation (MPE) [4] is motivated by problems in computational fluid dynamics and was later transferred to problems in circuit simulation [3]. Here, no full Galerkin-projection is applied, i.e. the multiplication with $W^T$ from the left is not done. Instead, after having constructed $V$, the full state $x$ in (3.1) is replaced with $Vz$. Then, a numerical integration method is applied. In the case of, e. g. a backward differentiation formula (BDF) method, this leads to nonlinear equations of the form

$$\alpha q(Vz_l) + s_l + j(Vz_l) + Bu(t_l) = 0, \tag{3.5}$$

with integration coefficient $\alpha \in \mathbb{R}$ and a history term $s_l = s_l(z_0, \ldots, z_{l-1}) \in \mathbb{R}^n$. Clearly, this specifies $n \in \mathbb{N}$ equations for the $(r \ll n)$ unknowns $z_l$ at some time-point $t_l$.

In MPE this overdetermined system is reduced to a still overdetermined system of dimension $r \leq g < n$, by neglecting $n - g$ equations. Introducing a selection matrix $P_g \in \{0,1\}^{g \times n}$, this choice can be described by multiplying (3.5) with this selection matrix from the left:

$$\alpha P_g q(Vz_l) + P_g s_l + j(Vz_l) + P_g Bu(t_l) = 0. \tag{3.6}$$

Clearly, this selection corresponds to evaluating just $g \ll n$ entries of $q$ and $j$. So, a reduction of function evaluations is obtained. As the system (3.6) is still overdetermined, it is solved in the least squares sense. The remaining question is: How to select, i.e., how to construct the selection matrix $P_g$, which has exactly one non-zero entry per row and at most one non-zero per column.

In MPE the answer to this question is connected directly to the snapshot matrix and the most dominant state variables, i.e., components of $x(t)$. Recall that the principal subspace is a subspace of the manifold the solution resides in. First an optimal orthonormal basis of the manifold is constructed, then the principal subspace arises from truncating this basis. The selection $P_g$ can be interpreted as an optimal truncation to a subspace of dimension $g$ w.r.t. the canonical basis. More formally, $P_g$ is chosen such that it minimizes

$$\| \left(V^T P_g^T P_g V\right) - I_{r \times r} \|,$$

i.e., even after canceling $n-g$ rows of $V$, the resulting matrix is close to being orthonormal. This optimization problem is solved in [4] by the so-called Greedy algorithm, a partly heuristic approach.

### Adapted POD

The adapted POD, introduced in [71], is based on the snapshots of the state $x(t)$, similar to what we have seen in MPE. Here, the flow of the POD approach is changed. From the SVD, see Section 3.2.1, there is not directly a matrix $V$, projecting to a lower-dimensional space, created. Instead the system is projected, Galerkin-like, on the full space spanned by $U$, i.e., without truncation. However, the projection is done with a $U$, scaled by the singular values, i.e., with

$$K = U \cdot \operatorname{diag}(\sigma_1, \dots, \sigma_n).$$

In this way, the dynamical system (3.1) is transformed to the, still full dimensional, problem:

$$\frac{d}{dt}[K^T q(Kw(t))] + K^T j(Kw(t)) + K^T Bu(t) = 0; \;\; y(t) = L^T K w(t),$$

where $w$ are the new coordinates of $x$ in the basis $K$. From that point on, $K$ and $K^T$ are treated in different ways. Both are approximated by matrices that coincide with $K$ and $K^T$, in all but $n-r$ and $n-g$ columns respectively which are set to $0 \in \mathbb{R}^n$. Technically, this is described with the help of selection matrices $P_r \in \{0,1\}^{r\times n}$ and $P_g \in \{0,1\}^{g\times n}$,

$$K \approx K P_r^T P_r \quad \text{and} \quad K^T \approx K^T P_g^T P_g.$$

The decision, which columns in $K$ and $K^T$ to keep and which to set to zero is based on the Euclidean norm of the columns. Since $K$ is the orthogonal matrix $U$ scaled with the singular values, $P_r$ will naturally choose the first $r$ columns. For $P_g$ no such conclusion can be drawn.

The final steps of the adapted POD are, to further approximate $K^T \approx P_r^T P_r K^T P_g^T P_g$, insert this, followed by some relabellings and finally, canceling the scaling with the singular values. So, we arrive at

$$\frac{d}{dt}\left[W_{r,g}\bar{q}(Vz)\right] + W_{r,g}\bar{j}(Vz) + \widehat{B}u(t) = 0,$$

with $U_r^T = P_r U^T = V$, $\bar{q}(\cdot) = P_g q(\cdot)$, $\bar{j}(\cdot) = P_g j(Vz)$, $W_{r,g} = V^T P_g^T \in \mathbb{R}^{r\times g}$ and $\widehat{B} = V^T B$. As in the MPE approach, $q$ and $j$, expensive in evaluation, are replaced by $\bar{q}$ and $\bar{j}$, consisting of just a small subset, and, hence are cheaper to evaluate.

### Discrete Empirical Interpolation

Another recent adaption of POD is motivated from the numerical solution of partial differential equations (PDEs), where, after discretization in space, the dynamical system exhibits a special structure. Discrete empirical interpolation method (DEIM), introduced in [11], can also be used for general nonlinear problems.

As in the two modifications of POD described above, finally, the nonlinear functions are replaced by a selective evaluation. In contrast to MPE and Adapted POD, DEIM is based on snapshots of the functions itself ("a dual approach") and not on those of the states $x$. In this way, the actual MOR scheme, i. e. the search for a reduced subspace to project the system on is decoupled from the restriction of the nonlinear functions.

The basic idea of DEIM is to represent a nonlinear function $f : \mathbb{R}^n \to \mathbb{R}^n$ approximately on a lower dimensional subspace, say, spanned by the columns of some matrix $U \in \mathbb{R}^{n\times g}$:

$$f(x) \approx Uc(x). \tag{3.7}$$

Equality cannot be achieved in general, as (3.7) defines an overdetermined system for the coefficients $c(x) \in \mathbb{R}^g$. Hence, only $g$ rows are chosen, which can, again by the help of a selection matrix $P_g \in \{0,1\}^{g\times n}$, be described by

$$P_g f(x) = (P_g U)c(x).$$

Note, that, given a regular matrix $(P_g U)$, it follows from (3.7):

$$f(x) \approx U\left(P_g U\right)^{-1} P_g f(x), \tag{3.8}$$

where again $P_g$ applied to $f(\cdot)$ means, that only $g$ out of the $n$ components of $f(\cdot)$ have to be evaluated. Clearly, $U\left(P_g U\right)^{-1}$ is constant and can be precalculated. Incorporating the POD ansatz to approximate $x \approx Vz$, finally we get a reduced dynamical system (3.2) with

$$\widehat{q}(z(t)) = \widehat{W}_q\, \bar{q}(z(t)), \quad \widehat{j}(z(t)) = \widehat{W}_j\, \bar{j}(z(t)), \quad \widehat{B} = V^T B, \quad \widehat{L} = V^T L,$$

with $\widehat{W}_q = V^T U_q \left(P_{g1}^q U_q\right)^{-1}$ and $\bar{q}(\cdot) = P_{g1}^q q(\cdot)$, $\widehat{W}_j = V^T U_j \left(P_{g2}^j U_j\right)^{-1}$ and $\bar{j}(\cdot) = P_{g2}^j j(\cdot)$.

The question open here is, how to determine $U$ and $P_g$. In DEIM, the former is constructed from a SVD on a matrix $F = (f(x_1), \ldots, f(x_K)) \in \mathbb{R}^{n\times K}$ of function snapshots, which can be taken, e. g. at time points $t_1, \ldots, t_K$ during the benchmark simulation. Consequently, $U$ consists of the $g$ most dominant left singular vectors of $F$.

The construction of $P_g$, i. e., the decision, which components of $f$ are dominant is the core of DEIM. With $U_l$ denoting the first $l$ columns of $U$ and with $P_l \in \{0,1\}^{l\times n}$ selecting $l$ columns from $U_l$, the algorithm incrementally computes the residual

$$r_{l+1} = u_{l+1} - U_l \left(P_l U_l\right)^{-1} P_l u_{l+1} \quad \text{for } l = 1, \ldots, g.$$

$P_1$ selects the entry of $U_1$ with the largest absolute value. Then $P_{l+1}$, selects in addition to what $P_l$ has selected the column with the index where the residual $r_{l+1}$ has the largest absolute value. Setting up the selection matrix in this way guarantees that $P_g U$ in (3.8) is regular.

## 3.3 Trajectory piecewise linear techniques

Trajectory piecewise linear techniques [6, 54, 55, 64, 73] linearize the nonlinear system around a finite number of suitably selected states and approximate the nonlinear system by a piecewise linearization that is obtained from combining the (reduced) linearized systems via a weighting procedure.

Having $s$ states $x_0, \dots, x_{s-1}$ obtained from simulation of the original system on some finite time interval $[t_{\text{start}}, t_{\text{end}}]$, we linearize the original system around these states:

$$\frac{d}{dt}(q(x_i) + Q_i(x(t) - x_i)) = j(x_i) + J_i(x(t) - x_i) + Bu(t), \tag{3.9}$$

where $x_0$ is the initial state of the system and $Q_i$ and $J_i$ are the Jacobians of $q$ and $j$ evaluated at $x_i$. Since each of the linearizations approximates the nonlinear system in the neighborhood of the expansion point $x_i$, a model including all these linearizations could approximate the original system over a larger time interval and larger part of the state space. In [54] a weighting procedure is described to combine the models. The piecewise linear model becomes [1]

$$\frac{d}{dt}\left[\sum_{i=0}^{s-1} w_i(x)Q_i x\right] = \sum_{i=0}^{s-1} w_i(x)\Big(j(x_i) + J_i(x - x_i)\Big) + Bu(t),$$

where $w_i(x)$ are state-dependent weights. Evaluation of weights has to be cheap. A typical choice is to let $w_i(x)$ be large for $x = x(t)$ close to $x_i$, and small otherwise, but other and more advanced schemes are also available [54].

Simulation of a piecewise linearized system may already be faster than simulation of the original nonlinear system. However, the linearized system can be reduced by using model order techniques for linear systems.

The main difference between a linear MOR and the nonlinear MOR approach TPWL is that the latter introduces in addition to the application of a linear MOR technique the selection of linearization points to get linear problems and the weighting of the linear sub-models to recover the global nonlinear behavior. In the following we present the steps in more details.

### 3.3.1 Selection of linearization points

The model extraction basically needs the solution of the full nonlinear system once. In [54] a fast extraction method is proposed, but we will not give details here.

The TPWL-scheme is based on deciding when to add a linear substitute for the non-linear problem automatically during simulation of the latter. Again there are several

[1] Note that $\frac{d}{dt}(q(x_i) - Q_i x_i) \equiv 0$.

alternatives. Rewieński [54] proposes to check at each accepted time point $t$ during simulation for the relative distance of the current state $x$ of the nonlinear problem to all existing linearization states $x_0, \ldots, x_{i-1}$. If the minimum is equal to or greater than some threshold $\delta > 0$, i.e.

$$\min_{0 \leq j \leq i-1} \left( \frac{\|x - x_j\|_\infty}{\|x_j\|_\infty} \right) \geq \delta, \tag{3.10}$$

then $x$ becomes the $(i+1)$th linearization point. Accordingly, a new linear model, from linearizing around $x$ is added to the collection. To our experience as we will show later, the choice of $\delta$ already marks a critical point in the process. Rewieński suggests to first calculate the steady state $x_T$ of the linear system that arises from linearizing the nonlinear system at the DC-solution $x_0$ and then setting $\delta = \frac{d}{10}$ where

$$d = \frac{\|x_T - x_0\|_\infty}{\|x_0\|_\infty} \quad (d = \|x_T\|_\infty \text{ if } x_0 = 0). \tag{3.11}$$

Choosing linearization points according to the criterion (3.10) essentially bases the decision on the slope of the trajectory the training is done on and not on the quality of the linear substitute model w.r.t. the nonlinear system. In Voß [73] the mismatch of nonlinear and linear system motivates the creation of a new linearization point and an additional linear model: as at each time point during training both the nonlinear and a currently active system are available, the latter one is computed in parallel to the former one. If the difference of the two approximations to the true solution at a timepoint $t_n$ produced by the different models becomes too large, a new linear model is created from linearizing the nonlinear system around the state the system was in at the previous timepoint $t_{n-1}$.

Note that in both strategies a linear model around a state $x(t_{n-1})$ is constructed when the linear model arising from linearization around a former state does not extend to $x(t_{n-1})$. However, it is not guaranteed that this new model extends backward, i.e., is able to reproduce the situation encountered before $t_{n-1}$. This circumstance could have a negative effect in re-simulation where linear combinations of linear substitute systems are used to replace the full nonlinear system. That means during model extraction one deals with just one linear system in each situation but with combinations during further simulations.

### 3.3.2 Determination of weights

During the training the nonlinear functions have been fragmented into linear ones, each part reflecting certain aspects of the *parent function*. When using the substitute collection for simulation, one will naturally aim at having to deal with a combination of just a small number of linear sub-models. Hence, the weighting function has to have steep gradients to determine a few (in the ideal case just one) dominant linear models.

As in Rewieński's work we implemented a scheme that is depending on the absolute distance of a state to the linearization points. The importance of each single model is defined by

$$w_i(x) = e^{-\frac{\beta}{m}\cdot\|x-x_i\|_2}, \quad \text{with } m = \min_j \|x - x_j\|_2. \tag{3.12}$$

By the constant $\beta$ we can steer how abrupt the change of models is. In Rewieński [54] $\beta = 25$ is chosen. To guarantee a convex combination, the weights are normalized such that $\sum_i w_i(x) = 1$. Note, that in (3.12) the absolute distance $\|x - x_i\|_2$ is taken in the full space $\mathbb{R}^n$. When using the reduced order model (3.13) for simulation one has to prolongate $\tilde{x} \in \mathbb{R}^k$ to the full space. Computational costs could be reduced if this reprojection was not necessary, i.e. if we could measure the distances in the reduced space already. If the linearization points $x_0, \ldots, x_{s-1}$ are in the space spanned by the columns of $V$, it suffices to project them once to the reduced space and take the distances there, i.e. calculate $\|\tilde{x} - \tilde{x}_i\|$ instead (cf. [54]). In the cited reference, it is not stated if extra steps are taken to guarantee that the linearization states are contained in the reduced space. Adding the linearization states to $V$ after orthogonalizing them against the columns of $V$ could be an appropriate activity, probably increasing the dimension of the reduced space.

However, taking no extra steps and just projecting the linearization points to the reduced space to take the distances there can be very dangerous as we present in Section 4. Therefore, we strongly recommend to project the reduced space back to the full space for measuring the distance to the linearization points.

### 3.3.3 Reduction of linear sub-models

Basically, any MOR-technique for linear problems can be applied to the linear sub-models. In [54] Rewieński proposes the usage of Krylov-based reduction using the Arnoldi-method. Vasilyev, Rewieński and White [69] introduce balanced truncation to TPWL and Voß [73] uses Poor Man's TBR as linear MOR kernel.

The common basis of all these methods is that one restricts to a dominant subspace of the state-space. Suppose that the columns of $V \in \mathbb{R}^{n\times k}$ span this dominant subspace of the state-spaces of the linearized systems, and that $W \in \mathbb{R}^{n\times k}$ is the corresponding test matrix. Then a reduced order model for the piecewise-linearized system can be obtained as

$$\frac{d}{dt}\left[\sum_{i=0}^{s-1} w_i(V\tilde{x})\big(W^T Q_i V\tilde{x}\big)\right]$$
$$= \sum_{i=0}^{s-1} w_i(V\tilde{x})\Big(W^T j(x_i) + W^T J_i(V\tilde{x} - x_i)\Big) + W^T Bu. \tag{3.13}$$

Here, all linear sub-models are reduced by projection with the overall matrices $V$ and $W$. Besides the choice of the reduction scheme, as mentioned above, the construction of these matrices is a further degree of freedom. In [54] two different strategies are presented. A simple approach is to regard only the linear model that arises from linearizing around the starting value $x_0$, construct a reduced basis $V_0$ and an according test space $W_0$ for this linear model and set $V = V_0$ and $W = W_0$. Hereby one assumes that the characterization dominant vs. not dominant does not change dynamically. In a more sophisticated manner, one derives reduced bases and test spaces $V_i$, $W_i$, respectively, for each linear sub-model $i = 0, \ldots, s-1$ and constructs $V$ and $W$ from $\{V_0, \ldots, V_{s-1}\}$ and $\{W_0, \ldots, W_{s-1}\}$. For more details we refer to [54, 73].

### 3.3.4 Construction of reduced order basis

For constructing a reduced order basis we have to take into account the linear sub-models.

The simplest approach bases the reduction on the linear model that arises from linearization around the DC solution $x_0$ only. From the corresponding system's matrices $G_0, F_0, B$ (cf. (3.9)) a basis $\{v_1, \ldots, v_l\}$ for the $l$th order Krylov subspace using the Arnoldi algorithm might be constructed. Then the matrix $V = [v_1, \ldots, v_l, \tilde{v}_0] \in \mathbb{R}^{N\times(l+1)}$, where $\tilde{v}_0$ arises from orthonormalization of $x_0$ versus $v_1, \ldots, v_l$ can be taken as the projection matrix for Galerkin projection of the linear combination of the linear sub-models.

A second, extended, approach might take into account all linear sub-models. Here in a first step reduced order models for each single linear subsystem are constructed, which yields local reduced subspaces spanned by the columns of $V_0, \ldots, V_{s-1}$. In a second step an SVD is done on the aggregated matrix $V_{\text{agg}} = [V_0, x_0; \ldots; V_{s-1}, x_{s-1}]$. The final reduced subspace is then spanned by the dominating left basis vectors of SVD. Note that by this truncation it cannot be guaranteed that the linearization points are in the reduced space.

## 3.4 Balanced truncation in nonlinear MOR

The energy $L_c(x_0)$ that is needed to drive a system to a given state $x_0$ and the energy $L_o(x_0)$ the system provides to observe the state $x_0$ it is in are the main terms in balanced truncation. A system is called balanced if states that are hard to reach are also hard to observe and vice versa, i.e. $L_c(x)$ large implies $L_o(x)$ large. Truncation, i.e. reduced order modeling is then done by eliminating these states.

For linear problems $L_c$ and $L_o$ are connected directly, by means of algebraic calculation, to the reachability and observability gramians $P$ and $Q$, respectively.

Recall that these can be computed from Lyapunov equations, involving the system matrices $C, G, B, L$ of the linear system (3.3). Balancing is reached by transforming the state space such that $P$ and $Q$ are simultaneously diagonalized:

$$P = Q = \mathrm{diag}(\sigma_1, \ldots, \sigma_n)$$

with the so called Hankel singular values $\sigma_1, \ldots, \sigma_n$, which are the square roots of the eigenvalues of the product $PQ$. From the basis that arises from the transformation only those basis vectors that correspond to large Hankel singular values are kept. The main advantage of this approach is that there exists an a priori computable error bound for the truncated system.

In transferring balanced truncation to nonlinear problems, three main tracks can be recognized. Energy consideration is the common background for the three directions.

1. In the approach suggested in [25] the energy functions arise from solving Hamilton-Jacobi differential equations. Similar to the linear case, a state-space transformation is searched such that $L_c$ and $L_o$ are formulated as quadratic form with diagonal matrix. The magnitude of the entries is used to determine the truncation again. The transformation is now state dependent, and instead of singular values, we get singular value functions. As the Hamilton-Jacobi system has to be solved and the varying state-space transformations have to be computed, it is an open issue, how the theory could be applied in a computer environment.

2. In sliding interval balancing [72], the nonlinear problem is first linearized around a nominal trajectory, giving a linear time varying system like

$$\frac{d}{dt}x(t) = A(t)x(t) + Bx(t), \quad y(t) = C(t)x(t).$$

   At each state, finite time reachability and observability gramians are defined and approximated by truncated Taylor series expansion. Analytic calculations, basically the series expansions, connect the local balancing transformation smoothly. This necessary step is the limiting factor for this approach in circuit simulation.

3. Finally, balancing is also applied to bilinear systems arising from linearizations of nonlinear problems. Here the key tool are so called algebraic gramians arising from generalized Lypunov equations. However, no one-to-one connection between these gramians and the energy functions $L_c$, $L_o$ can be made, but rather they can serve to get approximative bounds for the aforementioned. Furthermore, convergence parameters have to be introduced to guarantee the solvability of the generalized Lyapunov equations. For further details we refer to [9, 14] and the references therein.

# Chapter 4

# Linear and nonlinear examples and numerical simulations

In the first part of this chapter we will introduce linear circuits and reduce them with techniques which have already been discussed in Chapter 2. Next we will study two more linear circuits and implement the direct approach with $\varepsilon$-embedding. The results of the reduction with the two scenarios introduced in Chapter 2 will be presented. In linear simulation a Bode magnitude plot of a transfer function plots the magnitude of $H(i\omega)$, in decibel, for a number of frequencies $\omega$ in the frequency domain of interest, see Section 4.2. If the transfer function of the original system can be evaluated at enough points $s = i\omega$ to produce an accurate Bode plot, the original frequency response can be compared with the frequency response of the reduced model. In the second part the nonlinear circuits such as an inverter chain and a nonlinear transmission line will be studied. We reduced the nonlinear circuits with two promising methods TPWL and POD, see Chapter 3, and also discuss some challenging problems and show the results.

## 4.1 Linear Circuits

**Example** 1- We choose an RLC ladder network shown in Figure 4.1. We set all the capacitances and inductances to the same value 1 while $R_1 = \frac{1}{2}$ and $R_2 = \frac{1}{5}$, see [45,61]. We arrange 201 nodes which gives us the order 401 for the mathematical model of the circuit.

If we write the standard MNA formulation, see Section 1.2, the linear dynamical system is derived. Applying the same notation as in Section 2.1, the system matrices will be as follows (for $K = 3$ for example):

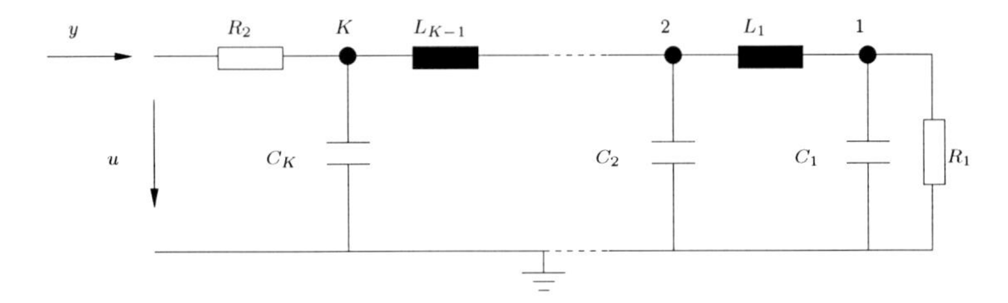


Figure 4.1: RLC Circuit of order $n = 2K - 1$, example 1.

$$C = I, \quad G = \begin{bmatrix} -2 & 0 & 0 & -1 & 0 \\ 0 & 0 & 0 & -1 & 1 \\ 0 & 0 & -5 & 0 & 1 \\ 1 & 1 & 0 & 0 & 0 \\ 0 & -1 & -1 & 0 & 0 \end{bmatrix}, \quad B = \begin{bmatrix} 0 \\ 0 \\ 5 \\ 0 \\ 0 \end{bmatrix},$$

$$L = \begin{bmatrix} 0 & 0 & -5 & 0 & 0 \end{bmatrix}, \quad D = 5. \tag{4.1}$$

The state variables are: $x_1$ the voltage across capacitance $C_1$; $x_2$ the voltage across capacitance $C_2$; ...; $x_K$ the voltage across capacitance $C_K$; $x_{K+1}$ the current through the inductance $L_1$; ...; $x_{2K-1}$, the current through the inductance $L_{K-1}$. In general the number of nodes $K$ is odd. The voltage $u$ and the current $y$ are input and output, respectively. Note that when the number of nodes is $K$ the order of the system becomes $n = 2K - 1$. In this test case we have an ODE instead of a DAE as $C = I$, see (4.1). The original transfer function is shown in Figure 4.2. The plot already illustrates how difficult it is to reduce this transfer function as many oscillations appear.

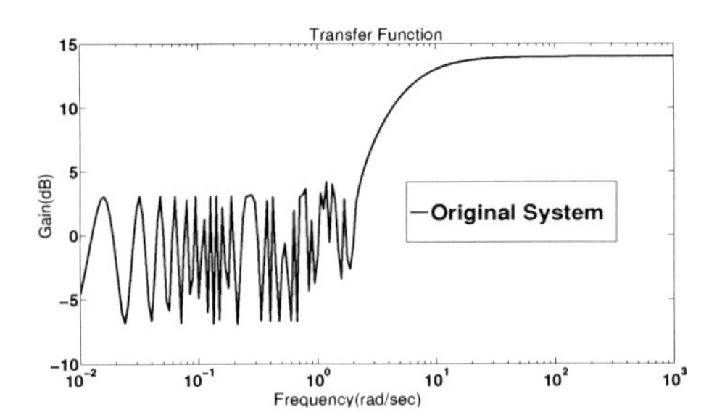


Figure 4.2: Original transfer function for the first example of Fig. 4.1, order $n = 401$. The frequency domain parameter $\omega$ varies between $10^{-2}$ to $10^3$.

**Example** 2- We use an RLC ladder network given in Figure 4.3 [57] for the second example.

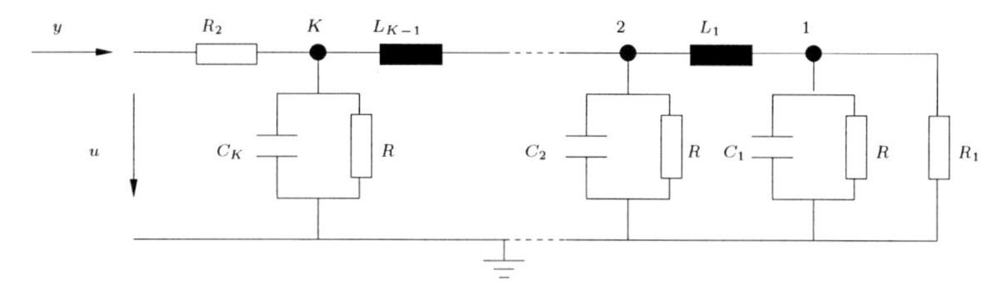


Figure 4.3: RLC Circuit of order $n = 2K - 1$, example 2.

The major difference to the previous example is that we introduced a resistor in parallel to the capacitors at each node connected to the ground. We set all the capacitances and inductances to the same value 1 while $R_1 = \frac{1}{2}$, $R_2 = \frac{1}{5}$ and $R = 1$. We choose 201 nodes which results in order 401 for the mathematical model of the circuit. Like the previous example we again derive a system of ODEs. The original transfer function of the second example is shown in Figure 4.4.

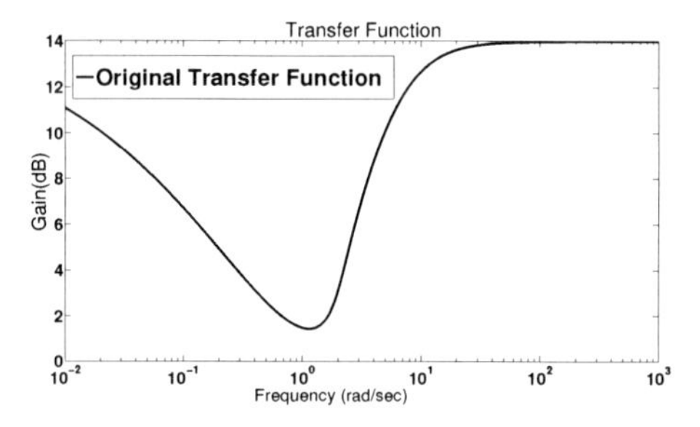


Figure 4.4: Original transfer function for the second example of Fig. 4.3, order $n = 401$. The frequency domain parameter $\omega$ varies between $10^{-2}$ to $10^3$.

The main reason for choosing these two examples is the behavior of Hankel singular values, see Figure 4.5. The Hankel singular values for the first example do not show any significant decay while in the second example we observe a rapid decay in the values.

Figures 4.6 and 4.7 show the absolute error between the transfer function of the full system and the transfer function of several reduced systems. The model is reduced by three linear techniques (PRIMA see Section 2.2.1, SPRIM see Section 2.2.1 and PMTBR see Section 2.2.2) for both examples.

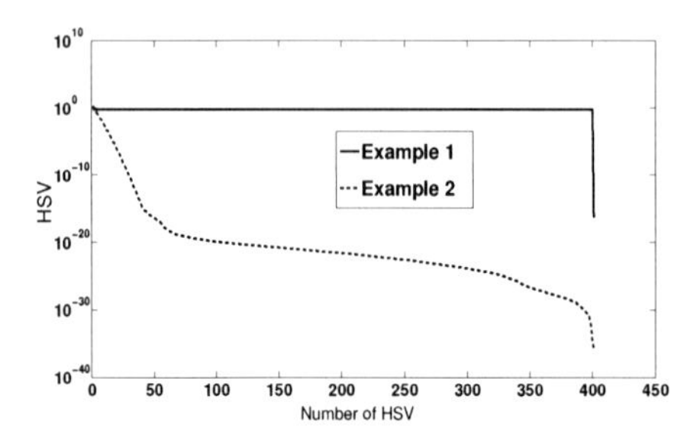


Figure 4.5: Hankel Singular Values for Example 1 and 2, (semi-logarithmic scale).

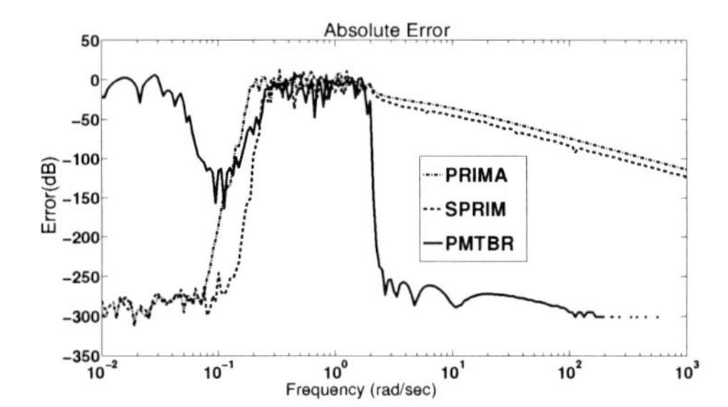


Figure 4.6: Error plot, the frequency domain parameter $\omega$ varies between $10^{-2}$ to $10^{3}$, Example 1.

In the example 1 we reduced the system from order $n = 401$ (number of nodes is $K = 201$) to order 34, which means that we reduced the system (in all three methods) by a factor of 10. The order of the reduction is relatively large in this case (thirty four) as the dynamical system is somehow stubborn for any reductions, see Figure 4.5. The price we should pay for the smaller system is too high as we lost a lot of information during the reduction and the error is becoming relatively large. As we expected, PRIMA and SPRIM in Figure 4.6 produced a reliable results close to the expansion point, in this case $s = 0$, but the error is immediately increasing for the rest of the oscillation part, see Figure 4.2, and then smoothly decreases for higher frequencies. In the first example the PMTBR match a bit worse for the low frequencies as the error decreases just for a short interval and immediately starts to increase again. But PMTBR also cannot cover the oscillation part of the transfer function although it resolves the higher frequencies well. The order in PMTBR results from a prescribed tolerance.

For the second example the SPRIM and PRIMA produced a nice match around the

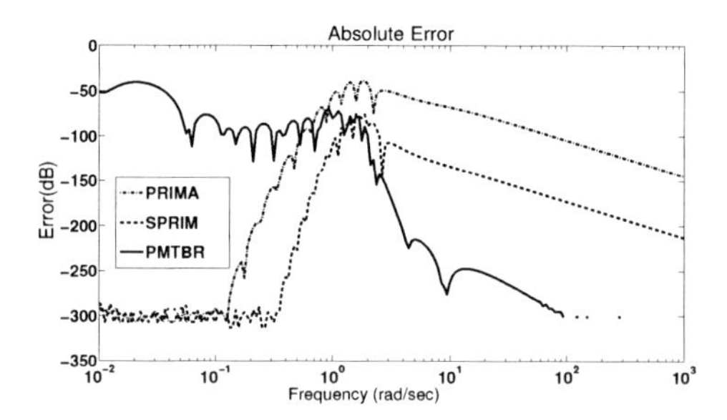


Figure 4.7: Error plot, the frequency domain parameter $\omega$ varies between $10^{-2}$ to $10^{3}$, Example 2.

expansion point $s = 0$ like the first example but for a larger interval, see Figure 4.7. The peaks of error both for PRIMA and SPRIM are around $-50$ and $-80$ dB, respectively, see Figure 4.7, which are much less than in example 1 where the peaks are around 0 dB for both PRIMA and SPRIM. We allowed the PMTBR to reduce the system by a factor of 20 in this case although we keep the order of the reduced system same as the first example for the PRIMA and SPRIM. Despite the lower dimension for reduced system the PMTBR produced much better results for this test case compared to the first example as the error starts from $-50$ dB and smoothly decreases for low frequencies and suddenly falls to $-300$ dB for larger frequencies.

As we expected, the SPRIM produces a better approximation than PRIMA, especially for the second example, since it matches twice as much moments. Although both methods have a good agreement around the expansion point $s = 0$, the error increases as we are far from the expansion point. Since the Hankel singular values for the first example do not decay, the PMTBR cannot produce an accurate model for low frequencies in that case. In the second example where the Hankel singular values rapidly decay the PMTBR produced a more reliable result with a better match. This shows that we cannot stick to one method for reduction in general and the method should be chosen depending on the circuit's behavior.

**Example** 3- We investigate a linear forced LC-oscillator, see [44], with a capacitor, a resistor and an inductor in series. As we want to extend this LC-oscillator to a scalable benchmark problem (both in differential part and algebraic part of a DAE model), we arrange $n$ resistors in parallel and $m$ capacitors in series, see Figure 4.8. A kind of sparse tableau modeling, cf. [34], has been used for this case, where the currents through the resistors are defined as unknowns. If we apply the modified nodal analysis (MNA) [31] to this case, then the equations of the algebraic constraints are not scalable any more. The values for capacitances and resistances are defined such that the total capacitance and the total resistance coincides for each $n$ and $m$. Moreover, all capacitances and resistances are chosen with the same parameters $C$ and $R$, respectively. An oscillation is forced by an input signal, since a current source

$$u(t) = I_0 \sin\left(\frac{2\pi}{T}t\right),$$

with fixed amplitude $I_0 > 0$ and period $T > 0$, is added at node 1.

In the case of resistors in parallel, the state variables $x \in \mathbb{R}^{m+n+2}$ consist of the voltages $e$ at the nodes, the current $i_L$ traversing the inductor, the currents $i_R$ running through the resistors and the current $i_I$ from the external source. The used physical parameters are

$$L_0 = 10^{-6}\ \mathrm{H}, \qquad R_{\mathrm{eq}} = 10^4\ \Omega, \qquad C_{\mathrm{eq}} = 10^{-9}\ \mathrm{F},$$

where $n = 300$ and $m = 300$ represent the numbers of the resistors and capacitors, respectively. For $i = 1, \dots, n$ and $j = 1, \dots, m$ each resistors and capacitors have the following value:

$$R_i = n10^4\ \Omega, \qquad C_j = m10^{-9}\ \mathrm{F}.$$

The matrix notation of the used sparse tableau analysis reads a DAE:

$$\begin{bmatrix} A_C C A_C^\top & 0 & 0 & 0 \\ 0 & L & 0 & 0 \\ 0 & 0 & 0 & 0 \\ 0 & 0 & 0 & 0 \end{bmatrix} \frac{d}{dt} \begin{bmatrix} e \\ i_L \\ i_R \\ i_I \end{bmatrix} + \begin{bmatrix} 0 & A_L & A_R & A_I \\ -A_L^\top & 0 & 0 & 0 \\ -GA_R^\top & 0 & I_{n\times n} & 0 \\ 0 & 0 & 0 & 1 \end{bmatrix} \begin{bmatrix} e \\ i_L \\ i_R \\ i_I \end{bmatrix} + \begin{bmatrix} 0 \\ 0 \\ 0 \\ -1 \end{bmatrix} u(t) = 0.$$

Now we apply the direct approach ($\varepsilon$-embedding) to example 3, for the details of the method see Section 2.4. MOR techniques can be applied within two scenarios: fixing a small $\varepsilon$ or performing the limit $\varepsilon \to 0$ in a parametric MOR, see Figures 2.1 and 2.2. We start with the first scenario with the variable $\varepsilon = 10^{-12}$ and the PMTBR is used as a reduction scheme for the ODE system. Figure 4.9 shows the transfer function both for the DAE and the ODE($\varepsilon$) and the reduced ODE both with fixed $\varepsilon$. The number in parentheses shows the order of the systems.

In Figure 4.10 the dashed line shows the absolute error between the DAE system and the ODE($\varepsilon$), we call it the direct approach error, while the errors of the full DAE to reduced ODE and full ODE($\varepsilon$) to reduced ODE are also shown. The results in Figure 4.10 are in agreement to Theorem 2.3 as well as the estimate (2.23) as the two systems show a reliable match for low frequencies and the error increases just for higher frequencies. At the low frequencies the error is dominated by the reduction error while at the higher frequencies it is dominated by the direct approach error. The solid line in Figure 4.10 shows the absolute error between the original DAE system and the reduced ODE with fixed $\varepsilon = 10^{-12}$. The underlying electric circuits represent benchmarks, since they are designed such that the MOR methods can be applied efficiently. Hence the PMTBR scheme produces reliable results in the test example, although the PMTBR is not an efficient reduction scheme for DAEs in general.

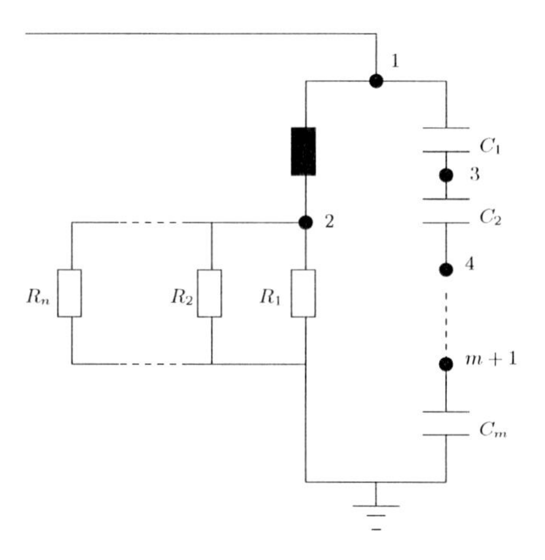


Figure 4.8: RLC circuit example 3.

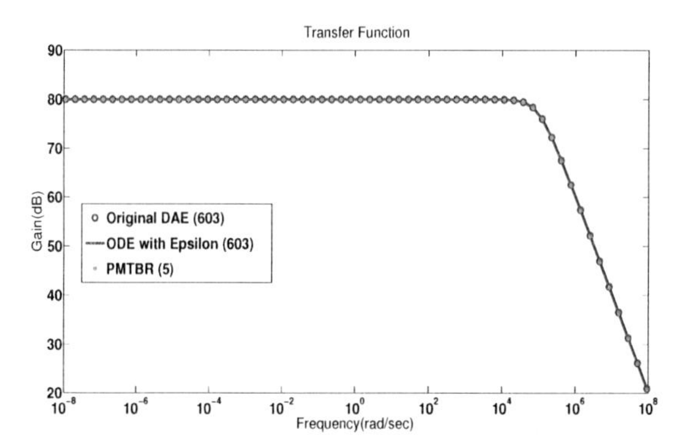


Figure 4.9: Original transfer function for DAE, ODE and reduced transfer function, example 3.

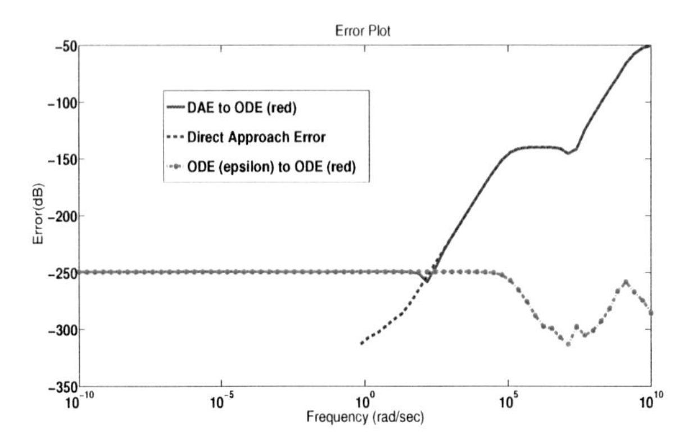


Figure 4.10: Corresponding error plot, reduction was achieved by PMTBR, example 3.

The second scenario with parametric MOR is studied now, see [41] for more details. Thereby, we apply the systems of Example 3 only. The limit $\varepsilon \to 0$ yields the desired result for the DAEs, see Figure 4.11. In Figure 4.11 we plot the reduced transfer functions with different values for $\varepsilon_0$. Hence in the same plot the sensitivities to the parameter $\varepsilon$ can be studied. We applied the PIMTAB, see Section 2.6, with $\varepsilon = 0$, $\varepsilon = 10^{-14}$, $\varepsilon = 10^{-10}$ and $\varepsilon = 10^{-7}$, see Figure 4.11.

We choose the same order as in the first scenario with the same numbers of resistors and capacitors $n = 300$ and $m = 300$, respectively, the order of the system is 603. In this case the parameter $\varepsilon$ has to be considered during the reduction. The absolute error of this parametric reduction scheme is shown in Figure 4.12. The transfer functions nearly coincide for $\varepsilon$ to zero or small enough ($\varepsilon = 10^{-14}$). For the case of larger $\varepsilon = 10^{-10}$ the results is almost as good as before except for very high frequencies, see Figure 4.12. In the case of relatively larger parameter, $\varepsilon = 10^{-7}$, we observe a larger error of the reduction scheme, since the error of the regularization becomes dominating. As the error for the higher frequencies goes above 20 dB we should say this value for the parameter is not acceptable. This is in agreement to Theorem 1 as well as the estimate (2.23).

The $\varepsilon$-embedding transforms a semi-explicit system of DAEs into a singularly perturbed system of ODEs. MOR methods for ODEs can be applied to the constructed system, where the parameter $\varepsilon$ is included. The input-output behavior of both DAE system and ODE system is described by respective transfer functions in frequency domain. We have shown that the transfer function of the singularly perturbed system of ODEs converges to the transfer function of the original DAEs in the limit of a vanishing parameter $\varepsilon$. We have presented numerical simulations, which produce good approximations using

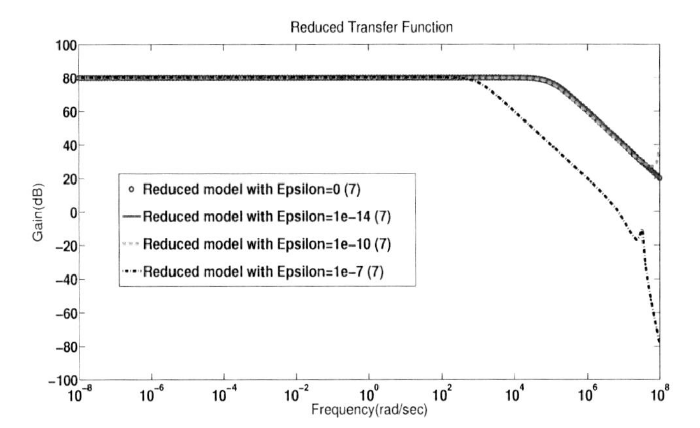


Figure 4.11: Reduced transfer function with second scenario (PIMTAB) in case of the parameters $\varepsilon = 0$, $\varepsilon = 10^{-14}$, $\varepsilon = 10^{-10}$ and $\varepsilon = 10^{-7}$, example 3.

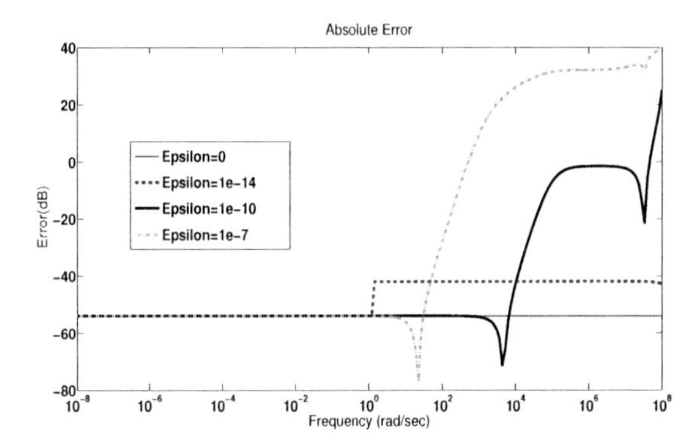


Figure 4.12: Corresponding error plot for $\varepsilon = 0$, $\varepsilon = 10^{-14}$, $\varepsilon = 10^{-10}$ and $\varepsilon = 10^{-7}$, example 3.

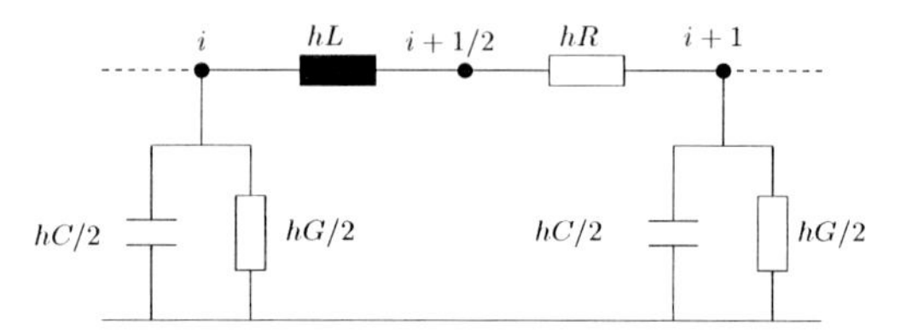


Figure 4.13: One cell of RLC model for transmission line, example 4.

the two approaches in case of a linear test example. In practice, we want to avoid to calculate the transfer function of the DAE or full ODE (the error is unknown) so the case of smaller $\varepsilon$ ends up with better results, but we should take into account that $\varepsilon$ is larger than the machin precision.

**Example** 4- We consider a substitute model of a transmission line (TL), see [30], which consists of $N$ cells. Each cell includes a capacitor, an inductor and two resistors, see Figure 4.13. This TL model represents a scalable benchmark problem (both in differential part and algebraic part but not separately), because we can select the number $N$ of cells. The state variables $x \in \mathbb{R}^{3N+3}$ consist of the voltages at the nodes, the currents traversing the inductances $L$ and the currents at the boundaries of the circuit. The used physical parameters are

$$C = 10^{-14}\ \text{F/m},\ L = 10^{-8}\ \text{H},\ R = 0.1\ \Omega/\text{m},\ G = 10\ \text{S/m}.$$

using the two approaches in case of a linear test example.

We apply modified nodal analysis, see [31], to the RLC circuit and then the state $x$ contains the $3N+3$ unknowns:

$$\begin{array}{cc} (V_0, V_1, \dots, V_N), & (I_{\frac{1}{2}}, I_{\frac{3}{2}}, \dots, I_{N-\frac{1}{2}}), \\ (V_{\frac{1}{2}}, V_{\frac{3}{2}}, \dots, V_{N-\frac{1}{2}}), & (I_0, I_N). \end{array}$$

We have $3N+3$ unknowns and only $3N+1$ equations. Thus two boundary conditions are necessary. Equations for the main nodes and the intermediate nodes in each cell are

$$\begin{array}{lllllll} \frac{h}{2}C\dot{V}_0 & + & \frac{h}{2}GV_0 & + & I_{\frac{1}{2}} & - & I_0 & = 0, \\ hC\dot{V}_i & + & hGV_i & + & I_{i+\frac{1}{2}} & - & I_{i-\frac{1}{2}} & = 0, \quad i = 1, \dots, N-1, \\ \frac{h}{2}C\dot{V}_N & + & \frac{h}{2}GV_N & + & I_N & - & I_{N-\frac{1}{2}} & = 0, \end{array}$$

$$\begin{array}{llll} -I_{i+\frac{1}{2}} & + & \frac{V_{i+1/2}-V_{i+1}}{hR} & = 0, \\ hL\dot{I}_{i+1/2} & + & (V_{i+1/2} - V_i) & = 0, \quad i = 0, 1, \dots, N-1, \end{array}$$

where the variable $h > 0$ represents a discretisation step size in space. We apply the boundary conditions

$$\begin{aligned} I_0 & - & u(t) &= 0, \\ L_1 \dot{I}_N & + & V_N &= 0. \end{aligned}$$

with $L_1 > 0$ and an independent current source $u$. Now a direct approach ($\varepsilon$-embedding) is used again, see Section 2.4. For the first simulation the variable $\varepsilon$ is fixed to $10^{-14}$ and $10^{-7}$, respectively, and the PMTBR is used as a reduction scheme for the ODE system. For all runs we selected the number of cells to $N = 300$, which results in the order $n = 903$ of the original system of DAEs (2.1). Figure 4.14 shows the transfer function both for the DAE and the ODE (including $\varepsilon$) and the reduced ODE with fixed $\varepsilon$. The number in parentheses shows the order of the systems.

Figure 4.15 illustrates the absolute error between the DAE system and the ODE (with $\varepsilon$). The direct approach error (dashed line), see Figure 4.15, is very small or zero for all three different values for $\varepsilon$ in low frequencies. So like the previous example at low frequencies the error is dominated by the reduction error and at higher frequencies by the direct approach error. The result of the error plot, see Figure 4.15, satisfies the inequality (2.23) as the error of the original DAE to ODE reduced is always less or equal the sum of the direct approach error and the linear reduction error. Also the two systems demonstrate a nice match for low frequencies and the error increases just for higher frequencies and then is smoothly decreasing for larger frequencies. We have a nearly perfect agreement between the original DAE and regularized DAE (i.e. the ODE with $\varepsilon$) when the $\varepsilon$ is small enough. It is clear that due to the small parameter $\varepsilon = 10^{-14}$ the error is below -200 dB for low frequencies and has a peak at -170 dB, but as we increase the parameter to $\varepsilon = 10^{-10}$ the error's peak increases to -100 dB and as we set $\varepsilon = 10^{-7}$ the error's peak occurs in -10 dB. As we do not obtain an acceptable error for the case $\varepsilon = 10^{-7}$, see Figure 4.15, this value for $\varepsilon$ is not accepted. In all cases the order of reduced model is 100 times smaller than the order of the original system and the reduced model is able to approximate the ODE system well.

Finally the second scenario with parametric MOR is studied. We apply the PIMTAB parametric MOR following [41]. The limit $\varepsilon \rightarrow 0$ gives the result for the reduced DAE, see Figure 4.16. The value in parentheses shows the order of the systems. We simulate again the TL model with $N = 300$ cells. We plot the transfer function for the parameters $\varepsilon = 0,\ 10^{-10},\ 10^{-7}$, see Figure 4.16. The error plot for the parametric reduction scheme is shown in Figure 4.17. The error plot shows an overall nice match for the case of $\varepsilon = 0, 10^{-14}$ and as the value for the parameter $\varepsilon$ increases, the accuracy of the method and of the reduction algorithm decrease. It is also important to mention that the order of the reduced system in the second scenario is nearly half of the one in the first scenario.

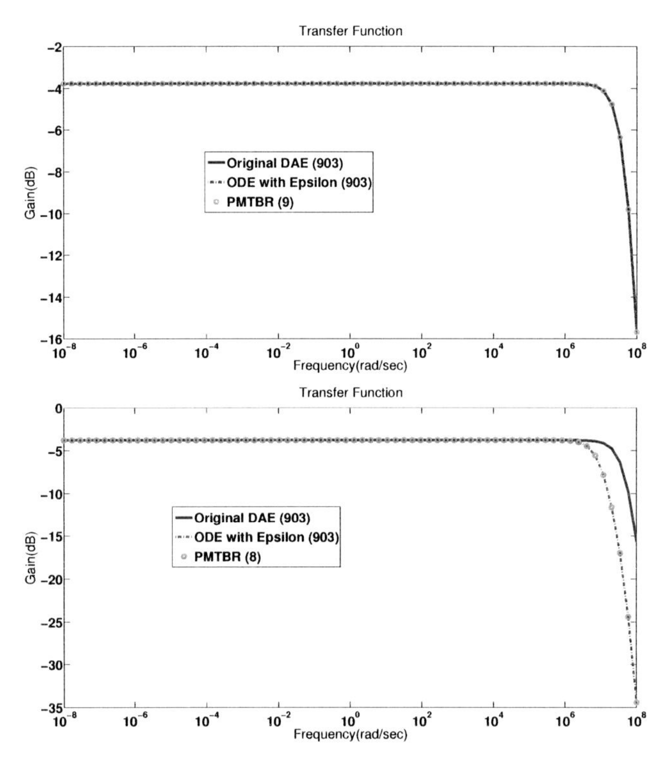


Figure 4.14: Original transfer function for DAE and ODE and reduced transfer function of PMTBR for $\varepsilon = 10^{-7}$ (down) and $\varepsilon = 10^{-14}$ (up), the frequency $\omega$ ranges from $10^{-8}$ to $10^{8}$, example 4.

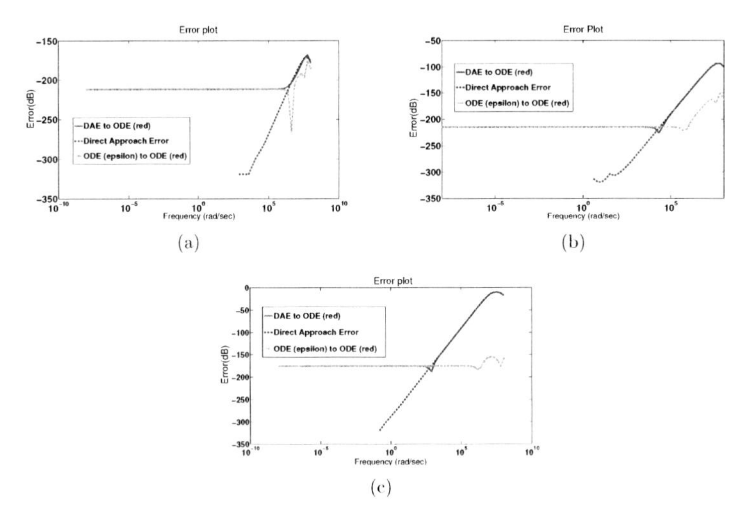


Figure 4.15: Absolute error plot for the $\varepsilon$-embedding and PMTBR methods in case of the parameters $\varepsilon = 10^{-14}$ (a), $\varepsilon = 10^{-10}$ (b) and $\varepsilon = 10^{-7}$ (c), the frequency $\omega$ ranges from $10^{-8}$ to $10^{8}$, example 4.

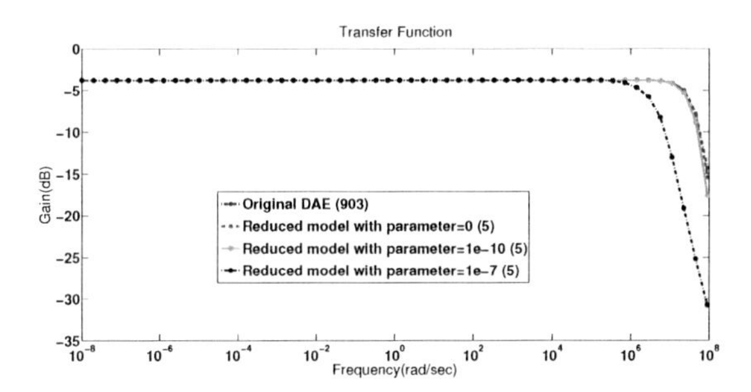


Figure 4.16: Original transfer function reduced with parametric scheme (second scenario) with different values for parameter $\varepsilon$, example 4.

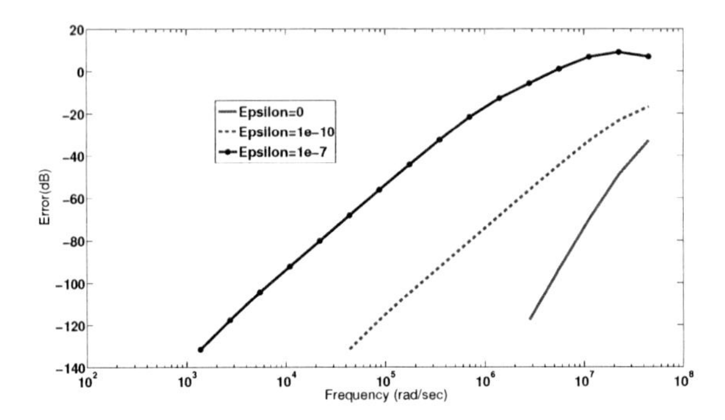


Figure 4.17: Absolute error plot for the $\varepsilon$-embedding, reduction carried out by parametric MOR (PIMTAB [42]), $\varepsilon = 0, 10^{-7}, 10^{-10}$, example 4.

## 4.2 Nonlinear Circuits

The TPWL and POD (and variants) are promising strategies in MOR for nonlinear circuit problems especially as they are up to now the only ones which can be implemented in a commercial circuit simulator. However both strategies suffer from high dependencies on heuristics. In case of TPWL the choices of linearization points and the weighting problem are the most important candidates for these heuristics. As we already mentioned in Chapter 3 there are two different strategies for choosing the linearization points, see [54, 73]. In the following we will show the results of TPWL for a nonlinear transmission line with these two strategies, see Section 3.3.1. The results of implementation of DEIM algorithm [11] with this transmission line is also presented. Finally the inverter chain with the TPWL will be studied. Test examples are the nonlinear transmission line (TL) model from Fig. 4.18 with $N = 100$ nodes, a problem of dimension 100, and the chain of $n = 100$ inverters from Fig. 4.19.

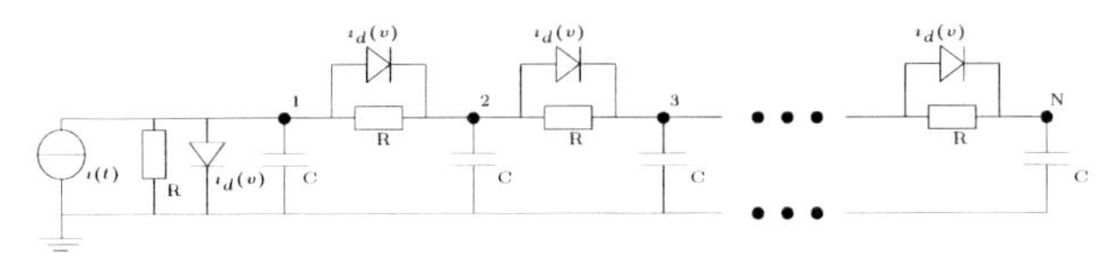

Figure 4.18: Nonlinear transmission line [54, 55].

### The nonlinear transmission line

The diodes in Figure 4.18 introduce the designated nonlinearity to the circuit, as the

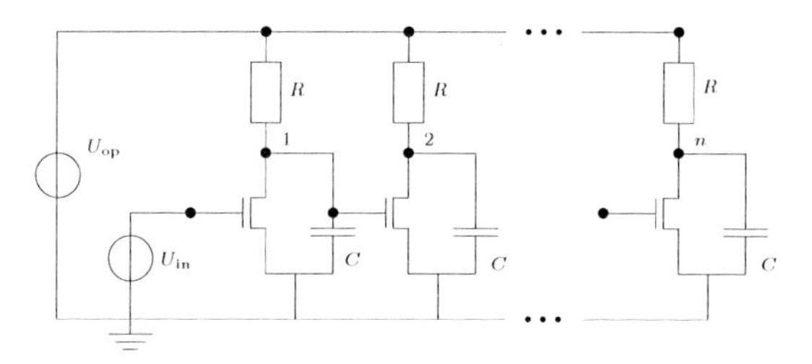


Figure 4.19: Inverter chain [73].

current $\imath_d$ traversing a diode is modeled by $\imath_d(v) = \exp(40 \cdot v) - 1$ where $v$ is the voltage drop between the diode's terminals. The resistors and capacitors contained in the model have unit resistance and capacitance ($R = C = 1$), respectively. The current source between node 1 and ground marks the input to the system $u(t) = \imath(t)$ and the output of the system is chosen to be the voltage at node 1: $y(t) = v_1(t)$.

Introducing the state vector $x = (v_1, \ldots, v_N)$, where $v_i$ describes the node voltage at node $i$, modified nodal analysis leads to the network equations:

$$\frac{dx}{dt} = f(x) + B \cdot u,$$
$$y = C^T \cdot x,$$

where $B = C = (1, 0, \ldots, 0)^T$ and $f : \mathbb{R}^N \to \mathbb{R}^N$ with

$$f(x) = \begin{pmatrix} -2 & 1 & & & \\ 1 & -2 & 1 & & \\ & \ddots & \ddots & \ddots & \\ & & 1 & -2 & 1 \\ & & & 1 & -1 \end{pmatrix} \cdot x + \begin{pmatrix} 2 - \exp(40x_1) - \exp(40(x_1 - x_2)) \\ \exp(40(x_1 - x_2)) - \exp(40(x_2 - x_3)) \\ \vdots \\ \exp(40(x_{N-2} - x_{N-1})) - \exp(40(x_{N-1} - x_N)) \\ \exp(40(x_{N-1} - x_N)) - 1 \end{pmatrix}.$$

### The inverter chain

In the inverter chain (Figure 4.19), the nonlinearity is introduced by the MOSFET-transistors. Basically, in a MOSFET transistor the current from drain to source is controlled by the gate-drain and gate-source voltage drops. Hence, the easiest way to model this element is to regard it as a voltage controlled current source and assume the leakage currents from gate to drain and gate to source to be zero:

$$\imath_{\text{ds}} = k \cdot l(u_g, u_d, u_s),$$

with $l(u_g, u_d, u_s) = \max(u_g - u_s - U_{\text{thres}}, 0)^2 - \max(u_g - u_d - U_{\text{thres}}, 0)^2$, where the threshold voltage $U_{\text{thres}} = 1$ and the constant $k = 2 \cdot 10^{-4}$ is chosen. For a more detailed definition of the corresponding network equations we refer to [45, 73].

### The simulation of nonlinear TL

Rewieński [54] uses this model Figure 4.18 to demonstrate the behavior of the automatic model extraction and the robustness of the TPWL-model w.r.t. input signals that differ from the training input signals. There, the simple approach for constructing the overall reduced subspace, i.e. just taking the reduction coming from the linear model that arises from linearization around the starting value is used. Additional linearization points are chosen according to the relative distance (3.10) to existing ones. No specification for $\delta$ therein is given. The Arnoldi-process to match $l = 10$ moments is used for reduction. Furthermore, the weighting procedure (3.12) is chosen with $\beta = 25$.

With these settings, five linear models are extracted automatically and the re-simulation shows a very good match even for differing input signals. With the same training- and simulation-input we try to reproduce these results. For the training the shifted Heaviside step function

$$\imath(t) = H(t-3) = \begin{cases} 0, & \text{if } t < 3 \\ 1, & \text{if } t \geq 3 \end{cases}$$

is used. Figure 4.20 shows both the short term (left) and the long term (right) response for all states, i.e. the node voltages . It can be seen that with increasing number the reaction of the nodes becomes weaker.

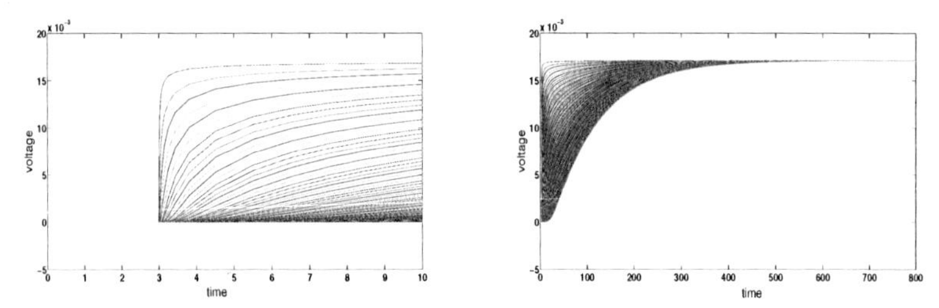


Figure 4.20: Transmission line: State response for all stages, left: short term simulations, right: long term simulations.

In Figure 4.21 we use Heaviside step function for resimulation also. It is clear that we have almost a perfect match with both the linear cores PMTBR and PRIMA. As the PMTBR and PRIMA produce almost the same results, for the next coming simulations we only use PRIMA for reducing the submodels.

As next test for re-simulation a different input signal

$$\imath(t) = 0.5 \cdot (1 + \cos(2\pi t/10)) \tag{4.2}$$

was applied to the reduced problem. In the following we give some details of the settings we used and the results we obtained.

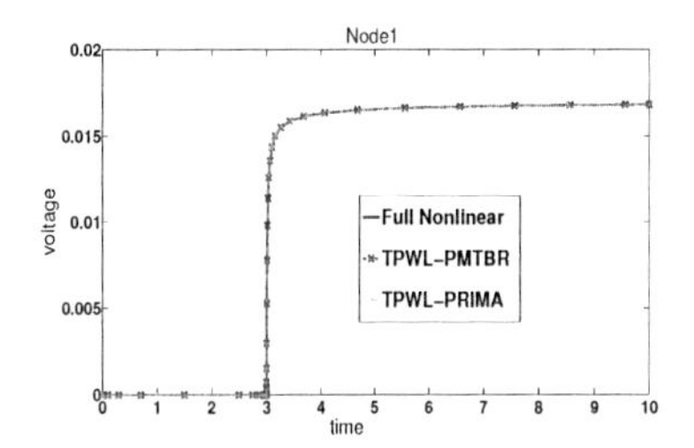


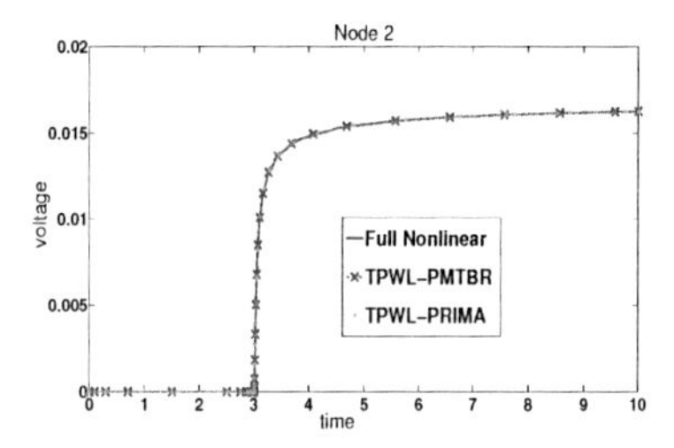


Figure 4.21: Transmission line: $N = 100$, as training input and resimulation signal we use Heaviside step function, Node 1 and 2.

- To get a guess for $\delta$, see (3.10), which is used for the selection of the linearization points, we run a simulation of an order reduced linear model that corresponds to the linearization around $x_0$ until $t_{\text{end}} = 300$ as at $t = 10$ steady state is not yet reached (see Figure 4.20). As the starting value satisfies $x_0 = 0$, $d$ and $\delta$ where chosen to be (see (3.11))

$$d = \|x_T\|_\infty \approx 0.0171 \quad \text{and} \quad \delta = 0.00171 \quad \text{with } T = 800.$$

- Both the simple and the extended approach for generating the reduced order basis were tested. For the latter attempt, the magnitude of the smallest singular value that was considered meaningful was 1% of the magnitude of the largest one.

Now we use two different strategies for choosing linearization points, see [54,73], and a new input signal, see (4.2), for resimulation. For both simulations automatic stepsize control with the same tolerances was used and all automatically extracted linear models were allowed to be used in the resimulation. With the Rewieński strategy [54] the method got 21 linearization points while with the Voß strategy [73] it is taken 31 linearization points. Finally the system was reduced to dimension 10.

We can clearly see that the model constructed in this way is not able to cope with the chosen new input signal, see Figure 4.22. Both strategies show a mismatch (jump) in the reduced model. However, this is not a matter of the reduction but of the process of selecting the linearized models. This observation can be made from having a look at Figure 4.24, which shows the result from replacing the full nonlinear with a full linear model, i.e., using the TPWL-procedure without reducing the linear submodels. The situation dramatically improves when only a subset of automatically extracted linear models is allowed to be used in resimulation, see Figure 4.23 when for the Rewieński approach we allowed submodels 1, 18 and for Voß the models 1, 18 and 20.

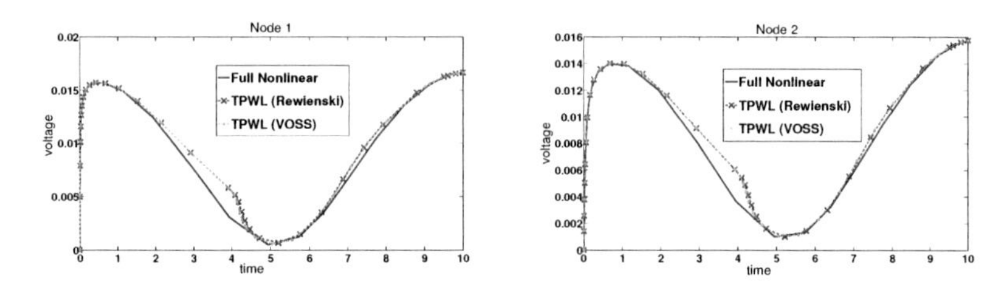


Figure 4.22: Transmission line: $N = 100$, as training input we use Heaviside step function and for resimulation signal we use Cosine function, the two MOR results are very close to each other, Node 1 and 2.

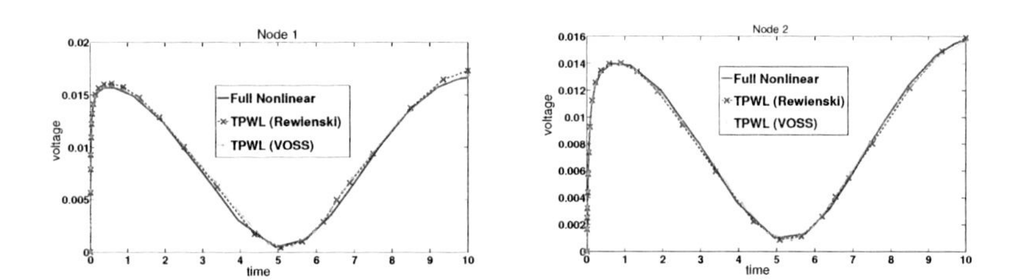


Figure 4.23: Transmission line: $N = 100$, as training input we use Heaviside step function and for resimulation signal we use Cosine function, selected linear models allowed, Node 1 and 2.

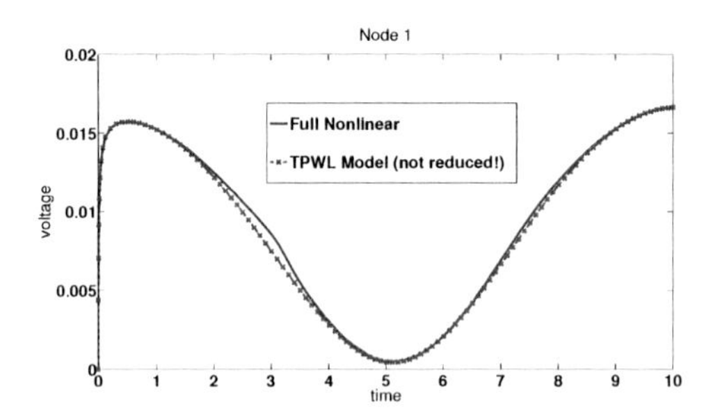


Figure 4.24: Transmission line: TPWL-re-simulation, no reduction, Node 1.

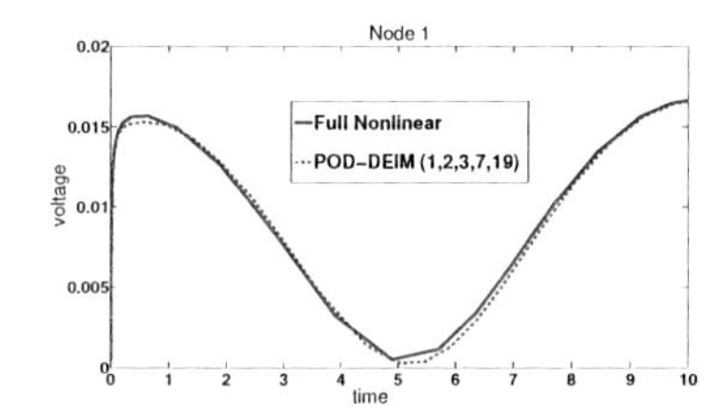


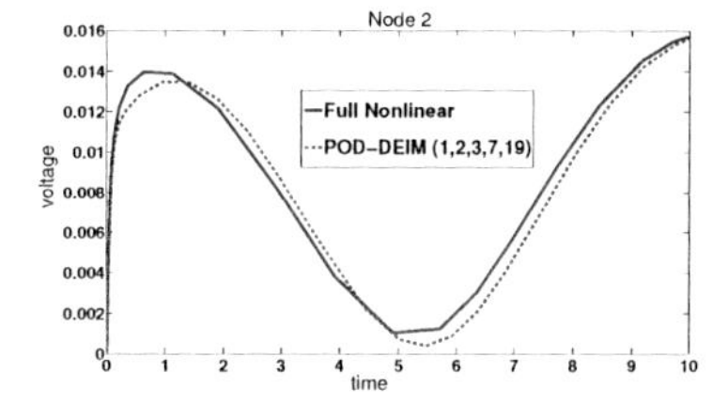


Figure 4.25: Transmission line: POD-DEIM with the cosine input, Node 1 and 2.

### Discrete Empirical Interpolation for the transmission line

We close this test example with an inspection on how the adapted POD approach [71] is behaving with the transmission line from Figure 4.18.
The DEIM algorithm, see Section 3.2.2, with sub-dimension $(1, 2, 3, 7, 19)$ , was implemented as well, see Figure 4.25. These sub-dimension numbers are the dominant components of a nonlinear function $f$, see (3.7).

In a first trial we use the same reduction parameter $r = 30$ and $g = 35$ for the dimension of the reduced state space and the dimension of the nonlinear element functions, respectively. With this setup the reduced order model is able to approximate the full nonlinear system as we see in Figure 4.25.

### The simulation results of the inverter chain

We apply only PRIMA and PMTBR as a linear core for TPWL. One of the partitions which is used inside the SPRIM algorithm is always of size 2 by 2 and the other part becomes larger as there is no inductor in the structure of the inverter chain. Therefore SPRIM is not reasonable to apply in this test case. The inverter chain constitutes a special class of circuit problems. Here a signal passes through the system, activating at each time-slot just a few elements and leaving the others latent. However, as the signal passes through, each element is active at some time and sleeping at some others. As in [45, 73], the training of the inverter chain during the TPWL model extraction was done with a single piecewise linear input voltage at $\bar{u}(t)$ (see also Figure 4.26 left), defined by

$$\bar{u}(0) = 0, \quad \bar{u}(5\text{ns}) = 0, \quad \bar{u}(10\text{ns}) = 5, \quad \bar{u}(15\text{ns}) = 5, \quad \bar{u}(17\text{ns}) = 0.$$

In Figure 4.27 and 4.28 we use PRIMA and PMTBR for reducing the linear subsystems. Figure 4.27 shows the state response of the inverter 4 both for the full nonlinear and the reduced version of it. The reduction with both linear cores shows a good match but we observe small oscillations which are caused by the linearization or the weighting procedure in both cases, see Figure 4.27. In Figure 4.28 the same scenario is repeated for inverter 52. Here we also observe the same oscillations for both TPWL with PRIMA and PMTBR kernel. On top of that the PMTBR also shows a little delay on inverter

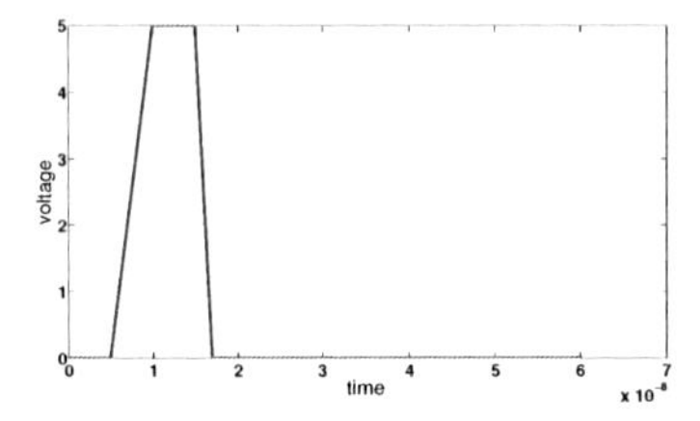


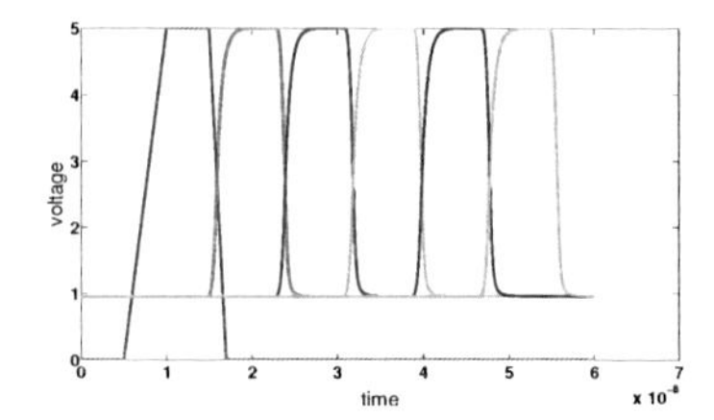


Figure 4.26: Inverter chain: training input (left) and state response (right, stages 1,20,40,60,80,100).

52 compared to the full nonlinear and PRIMA version. So for the next simulations we only stick to PRIMA as linear cores for reduction. Now the two pulses are used as

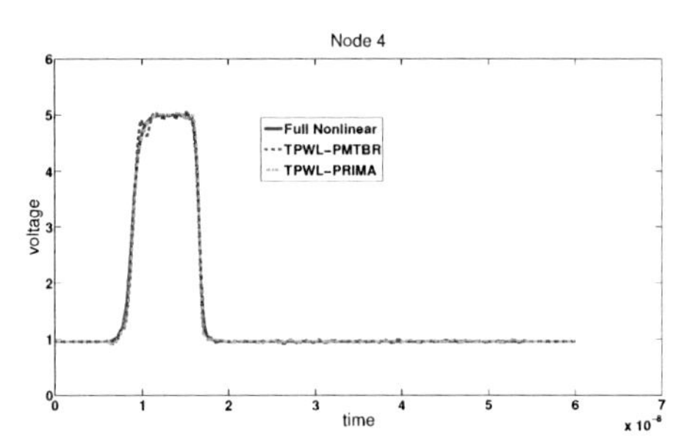


Figure 4.27: Inverter chain: TPWL-re-simulation, reduction to order 50, linear cores PRIMA and PMTBR, inverter 4.

input. In Figure 4.30 and 4.31 the voltage at inverters 70 and 80 is given. In both cases, the signal cannot be recovered correctly. In the latter case the pulse is even not recognized at all while we do not observe such a problem for the very beginning of the nodes, see Figures 4.29. At the moment we cannot state reasons for that. Obviously this is not caused by the reduction but by the linearization or the weighting procedure as we get similar results when turning off the reduction step.

The impact of broadening the input signal $u$ in (re-)simulation only can be seen in Figures 4.32 and 4.33 which displays the voltage at inverters 7 and 30. The signals are far away from the expected behavior. However, there seems to be a trend towards the situation that was encountered during the training. Indeed in Figure 4.33, at inverter 30 we find a time shifted version of the training signal instead of the wide input signal that has been applied now.

Finally, in Figures 4.34 and 4.35 the result of using the reduced model that arises from a training input $\bar{u}$ of given pulse width with a slightly tighter input signal $u$ is given

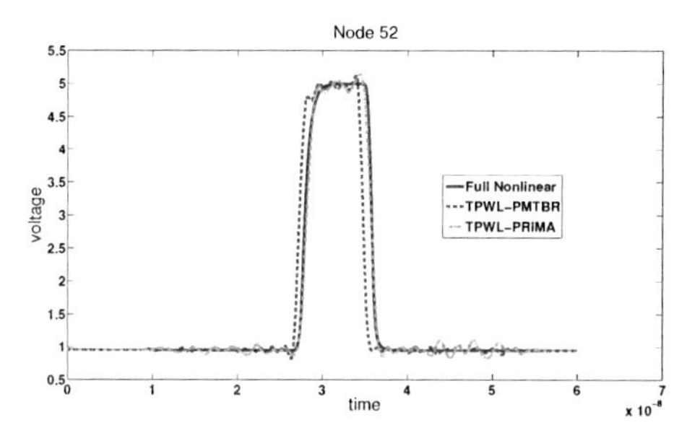


Figure 4.28: Inverter chain: TPWL-re-simulation, reduction to order 50, linear cores PRIMA and PMTBR, inverter 52.

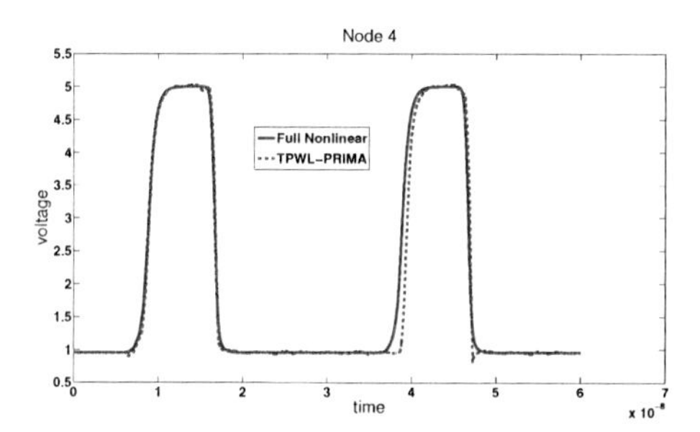


Figure 4.29: Inverter chain: TPWL-one pulse in training input and repeated pulse in re-simulation input, reduction to order 50, inverter 4.

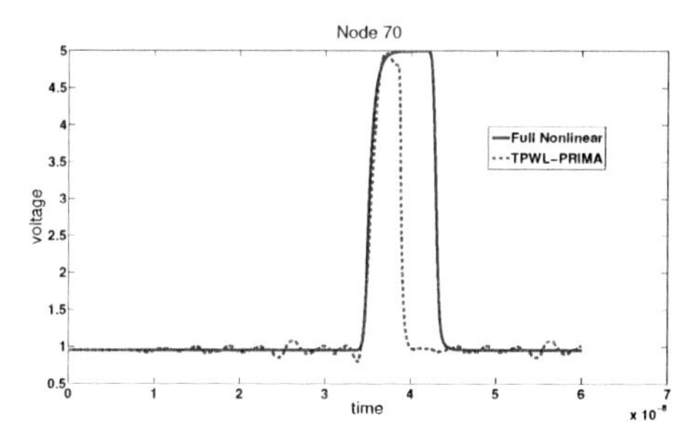


Figure 4.30: Inverter chain: TPWL-one pulse in training input and repeated pulse in re-simulation input, reduction to order 50, inverter 70.

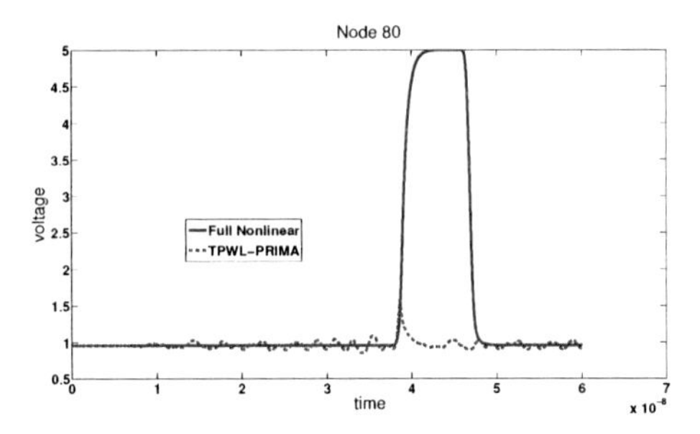


Figure 4.31: Inverter chain: TPWL-one pulse in training input and repeated pulse in re-simulation input, reduction to order 50, inverter 80.

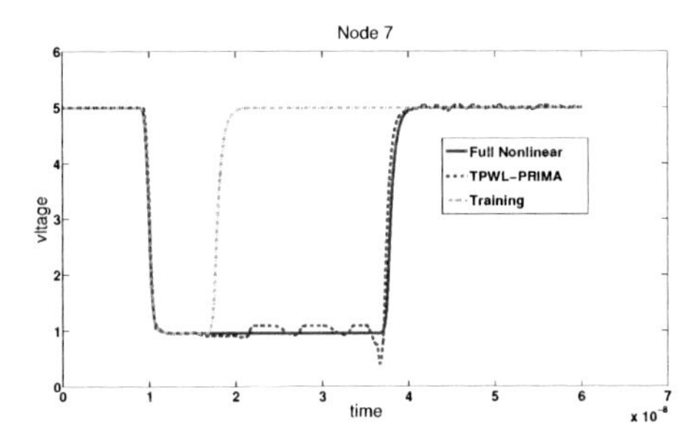


Figure 4.32: Inverter chain: TPWL-re-simulation, reduction to order 50, wider pulse, inverter 7.

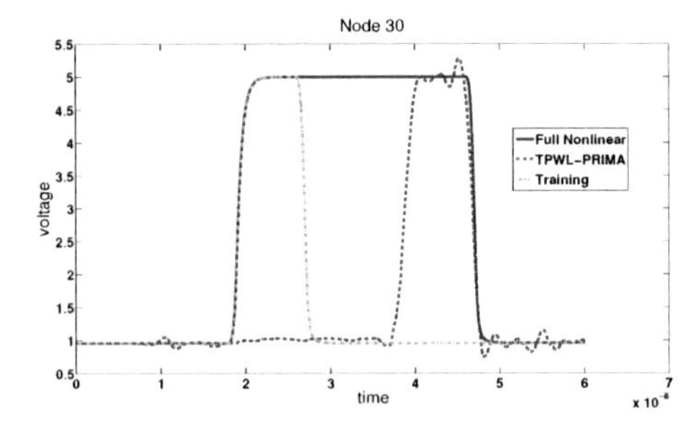


Figure 4.33: Inverter chain: TPWL-re-simulation, reduction to order 50, wider pulse, inverter 30.

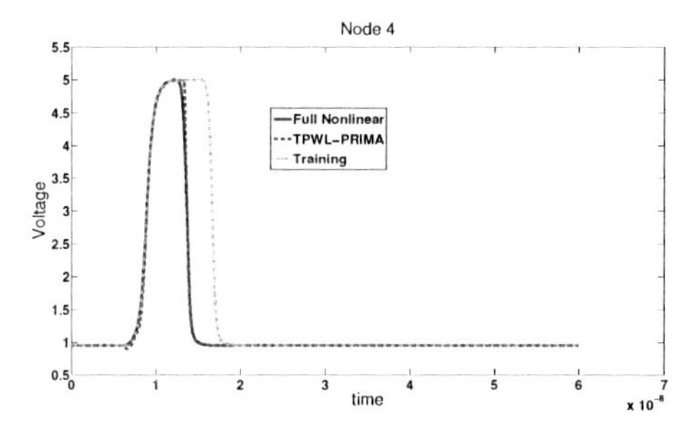


Figure 4.34: Inverter chain: TPWL-re-simulation, reduction to order 50, tighter impulse, inverter 6.

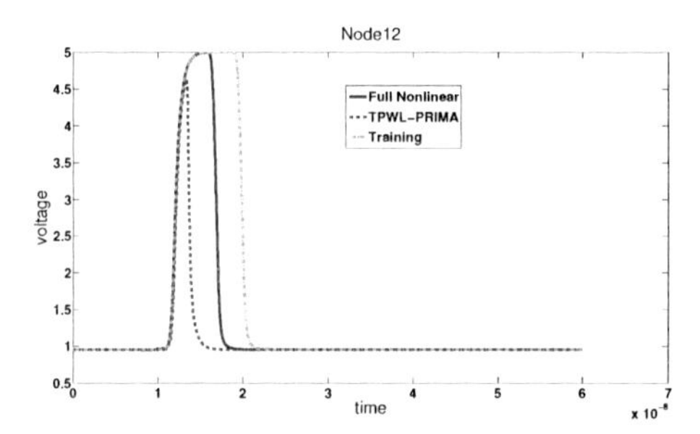


Figure 4.35: Inverter chain: TPWL-re-simulation, reduction to order 50, tighter impulse, inverter 12.

for the inverters 6 and 12, respectively. In the former the characteristic is reflected quite well. However, in the latter the output signal surprisingly seems to be a smaller pulse than the training signal, which is the qualitatively correct tendency but still do not match the full nonlinear problem . Having a closer look at how the inverter chain is modeled we see that the input voltage is applied at a floating node. This could give reasoning for the observed behavior. However, also the backward and forward validity of the linear models could be the reasons.

# Chapter 5

# Conclusions and outlook

We applied linear methods for reduction to some linear circuits. We studied a Krylov based classical method for reduction which is PRIMA and then compared it with recent methods such as SPRIM, adapted version of PRIMA which can preserve the structure, and PMTBR which combines the TBR method with projection techniques. We applied all three methods to two different test examples. Although the dynamical system for both examples yields an ODE all three methods are unable to completely cover the oscillation part of the transfer function in the first example. In contrast all the methods especially SPRIM and PMTBR produce a reliable reduction for the second example. Our observation shows the different behavior of the Hankel singular values of the two examples. This indicates that although the linear methods nowadays are well defined and tested we cannot stick to one method for reduction in general and the method should be chosen depending on the circuit's behavior.

The $\varepsilon$-embedding, i.e., the direct approach, is applied to a general linear system of DAEs via a singular value decomposition. Thereby, we obtain an approximating system of ODEs. Then, MOR techniques perform the reduction of the system of ODEs for the two scenarios of a fixed $\varepsilon$ or a variable parameter (parametric scheme). The presented approach enables the usage of MOR methods for ODEs. Most of the linear reduction schemes are designed and adopted for ODEs such as PMTBR or the spectral zeros preservation. We investigated a linear forced LC-oscillator and a linear transmission line model as test examples for the direct approach. Both test cases have been simulated successfully in both scenarios. The next step is to test the method on examples from industrial applications. Further investigations are necessary to apply the approach to nonlinear systems.

Examples of nonlinear systems arise in almost all applications in electrical engineering: the presence of a single transistor or diode already makes the circuit behavior nonlinear. More concrete examples are phase locked loops (PLLs), that usually contain voltage controlled oscillators (VCOs) and transmission lines with nonlinear elements such as diodes. Especially the latter can be good candidates for nonlinear model order reduction techniques: they usually have a limited number of inputs and outputs and exhibit a smooth behavior that is suitable for reduction. Other systems, such as inverter chains, show more digital behavior and are much more difficult for model reduction techniques, since the behavior cannot be described by a few dominant states. Here other techniques such as multirate time integration [63] may be better alternatives.

In contrast to the linear MOR where the main goal is to reduce the number of the states, here in nonlinear MOR the main objective is to construct a reduced model in the sense that it can be reused at much lower computational costs. Piecewise linearization in combination with linear MOR techniques in principle offers both reduction in the number of states and in simulation time. Also recent adaption to POD is motivated from the numerical solution of partial differential equations, where, after discretization in space, the dynamical system exhibits a special structure. The DEIM can however also be used for general nonlinear problems.

We tested a transmission line model and an inverter chain with TPWL and DEIM. However, both strategies suffer from high dependencies on heuristics. In case of TPWL the choice of linearization points and the weighting problem are the most important candidates for these heuristics. At the moment the TPWL is not robust enough.

A clear result for TPWL is that it is (much) more accurate and robust to determine the weights using state vectors in the original state space, i.e., reduced state vectors must be projected back to the original full state space before determining the weights. Numerical experiments confirmed that more accurate results are achieved in this way.

Another important issue is related to the reuse of piecewise-linear models. Since the piecewise-linear model is constructed using a single training input, this training input needs to be chosen carefully. Our experiments, however, indicate that even with an input close to the training input, simulation of the piecewise-linear models may become inaccurate. This also causes doubts on the robustness of the current linearization schemes.

# Bibliography


[1] A. C. Antoulas. *Approximation of large-scale Dynamical Systems, advances in design and control.* SIAM, 2005.

[2] U. M. Ascher and L. R. Petzold. *Computer Methods for Ordinary Differential Equations and Differential-Algebraic Equations.* SIAM, 1998.

[3] P. Astrid and A. Verhoeven. Application of least squares MPE technique in the reduced order modeling of electrical circuits. In *Proceedings of the 17th Int. Symp. MTNS*, pages 1980–1986, 2006.

[4] P. Astrid, S. Weiland, K. Willcox, and T. Backx. Missing point estimation in models described by proper orthogonal decomposition. *IEEE Transactions on Automatic Control*, 53(10):2237–2251, 2008.

[5] S. Azou and N. Munro. A new discrete impulse response gramian and its application to model reduction. *IEEE Transactions on Automatic Control*, 45(3):533–537, 2000.

[6] T. Bechtold, M. Striebel, K. Mohaghegh, and E. ter Maten. Nonlinear model order reduction in nanoelectronics: combination of POD and TPWL. In *PAMM (Proc. in Appl. Maths and Mechanics), special issue on Proceedings GAMM annual Meeting 2008*, pages 10057–10060, 2009.

[7] T. G. J. Beelen. *New algorithms for computing the Kronecker canonical form of a pencil with applications to systems and control theory.* PhD thesis, Eindhoven University of Technology, 1985.

[8] P. Benner. Advances in balancing-related model reduction for circuit simulation. In J. Roos and L. R. G. Costa, editors, *Scientific Computing in Electrical Engineering SCEE 2008*, volume 14, pages 469–482. Springer, 2010.

[9] P. Benner and T. Damm. Lyapunov equations, energy functionals and model order reduction. in preparation.

[10] G. Berkooz, P. Holmes, and J. Lumley. The proper orthogonal decomposition in the analysis of turbulent flows. *Annual Review of Fluid Mechanics*, 25:539–575, 1993.

[11] C. Chaturantabut and D. C. Sorensen. Discrete empirical interpolation for nonlinear model reduction. 2009.

[12] L. Chua, C. Desoer, and E. Kuh. *Linear and nonlinear circuits.* Mc Graw-Hill, New York, 1987.

[13] L. O. Chua and P. M. Lin. *Computer-Aided analysis for electronic circuits.* Prentice Hall, Englewood Cliffs, 1975.

[14] M. Condon and R. Ivanov. Nonlinear systems – algebraic gramians and model reduction. *COMPEL: The International Journal for Computation and Mathematics in Electrical and Electronic Engineering*, 24(1):202–219, 2005.

[15] L. Daniel, O. C. Siong, L. S. Chay, K. H. Lee, and J. White. A multiparameter moment-matching model-reduction approach for generating geometrically parameterized interconnect performance models. *IEEE Trans. CAD*, 23(5):678–693, 2004.

[16] C. A. Desoer and E. S. Kuh. *Basic circuit theory.* McGraw-Hill, New York, 1969. Chapter 12.

[17] P. V. Dooren. The generalized eigenstructure problem in linear system theory. *IEEE Trans. Aut. Contr.*, AC-26:111–129, 1981.

[18] P. Feldmann and R. W. Freund. Efficient linear circuit analysis by Padé approximation via the Lanczos process. *IEEE Transactions on Computer-Aided Design of Integrated Circuits and System*, 14(5):639–649, 1995.

[19] L. Feng. Parameter independent model order reduction. *Mathematics and Computers in Simulation*, 68(3):221–234, 2005.

[20] L. Feng and P. Benner. A robust algorithm for parametric model order reduction based on implicit moment matching. In *Proceedings in Applied Mathematics and Mechanics*, 2007.

[21] L. Feng, E. B. Rudnyi, and J. G. Korvink. Preserving the film coefficient as a parameter in the compact thermal model for fast electrothermal simulation. *IEEE Trans. on CAD of Integrated Circuits and Systems*, 24(12):1838–1847, 2005.

[22] R. Freund. SPRIM: structure preserving reduced-order interconnect macromodeling. In *Proc. ICCAD*, pages 80–87, 2004.

[23] R. W. Freund. Passive reduced-order models for interconnect simulation and their computation via Krylov-subspace algorithms. In *36th ACM/IEEE Design Automation Conference*, pages 195–200, 1999.

[24] R. W. Freund. *Model Order Reduction: Theory, Research Aspects and Applications*, chapter Structure-Preserving Model Order Reduction of RCL Circuit Equations, pages 49–73. Springer, 2008.

[25] K. Fujimoto and J. M. A. Scherpen. Singular value analysis and balanced realizations for nonlinear systems. In W. H. A. Schilders, H. A. van der Vorst, and J. Rommes, editors, *Model Order Reduction: Theory, Research Aspects and Applications*, pages 251–272. Springer Berlin Heidelberg, 2008.

[26] K. Gallivan, E. Grimme, and P. V. Dooren. Asymptotic waveform evaluation via a Lanczos method. In *Proceedings of the 33rd Conference on Decision and Control*, volume 1, pages 443–448, 1994.

[27] C. W. Gear. Differential-algebraic equations index transformations. *J. Sci. Stat. Comput.*, 9:39–47, 1988.

[28] E. Griepentrog and R. März. Differential-algebraic equations and their numerical treatment. *Teubner Texte zur Mathematik*, 88.BSB B. G, 1986.

[29] E. J. Grimme. *Krylov Projection Methods for Model Reduction.* PhD thesis, University of Illinois at Urbana-Champaign, IL, 1997.

[30] M. Günther. *Partielle differential-algebraische Systeme in der numerischen Zeitbereichsanalyse elektrischer Schaltungen.* Fortschritt-Berichte VDI Reihe 20. VDI, Düsseldorf, 2001.

[31] M. Günther and U. Feldmann. CAD based electric circuit modeling in industry I: mathematical structure and index of network equations. volume 8 of *Surv. Math. Ind.*, pages 97–129, 1999.

[32] M. Günther, U. Feldmann, and E. J. W. ter Maten. Modelling and discretization of circuit problems. In P. G. Ciarlet, W. H. A. Schilders, and E. J. W. ter Maten, editors, *Numerical Methods in Electromagnetics*, volume XIII of *Handbook of Numerical Analysis*, pages 523–659. Elsevier, North Holland, 2005.

[33] P. Gunupudi and M. Nakhla. Multi-dimensional model reduction of VLSI interconnects. In *Proceedings of IEEE Custom Intergrated Circuits Conference*, pages 499–502, 2000.

[34] G. Hachtel, R. Brayton, and F. Gustavson. The sparse tableau approach to network analysis and design. *Circuit Theory, IEEE Transactions on*, 18(1):101–113, Jan 1971.

[35] E. Hairer and G. Wanner. *Solving Ordinary Differential Equations II: Stiff and Differential-Algebraic Problems.* Springer, Berlin, 2nd edition, 1996.

[36] C. W. Ho, A. Ruehli, and P. Brennan. The modified nodal approach to network analysis. *Circuits and Systems, IEEE Transactions on*, 22(6):504–509, Jun 1975.

[37] P. Holmes, J. Lumley, and G. Berkooz. *Turbulence, Coherent Structures, Dynamical Systems and Symmetry.* Cambrige University Press, Cambrige, UK, 1996.

[38] J.-R. Li. *Model Reduction of Large Linear Systems via Low Rank System Gramians*. PhD thesis, Massachusetts Institute of Technology, Cambridge, MA, 2000.

[39] X. Li, P. Li, and L. T. Pileggi. Parameterized interconnect order reduction with explicit-and-implicit multi-parameter moment matching for inter/intra-die variations. *IEEE/ACM ICCAD*, pages 806–812, 2005.

[40] Y. Li, Z. Bai, and Y. Su. A two-directional Arnoldi process and its application to parametric model order reduction. *Journal of computational and applied mathematics*, 226(1):10–21, 2009.

[41] Y. Li, Z. Bai, Y. Su, and X. Zeng. Parameterized model order reduction via a two-directional Arnoldi process. In *IEEE/ACM international conference on computer-aided design*, pages 868–873, 2007.

[42] Y.-T. Li, Z. Bai, Y. Su, and X. Zeng. Model order reduction of parameterized interconnect networks via a two-directional Arnoldi process. *IEEE Trans. on CAD of Integrated Circuits and Systems*, 27(9):1571–1582, 2008.

[43] M. Loève. *Probability Theory*. Van Nostrand, 1955.

[44] K. Mohaghegh, R. Pulch, M. Striebel, and J. ter Maten. Model order reduction for semi-explicit systems of differential algebraic equations. In I. Troch and F. Breitenecker, editors, *Proceedings MATHMOD 09 Vienna - Full Papers CD Volumes*, pages 1256–1265. Springer, 2009.

[45] K. Mohaghegh, M. Striebel, J. ter Maten, and R. Pulch. Nonlinear model order reduction based on trajectory piecewise linear approach: comparing different linear cores. In J. Roos and L. R. G. Costa, editors, *Scientific Computing in Electrical Engineering SCEE 2008*, volume 14, pages 563–570. Springer, 2010.

[46] B. C. Moore. Principal component analysis in linear systems: controllability, observability, and model reduction. *IEEE Transactions on automatic control*, 26(1):17–32, 1981.

[47] A. Odabasioglu, M. Celik, and L. Paganini. Prima: Passive reduced-order interconnect macromodeling algorithm. *IEEE TCAD of Integ. Circuits and Systems*, 17(8):645–654, 1998.

[48] J. Phillips and L. M. Silveira. Poor's man TBR: a simple model reduction scheme. In *DATE*, volume 2, pages 938–943, 2004.

[49] J. R. Phillips, L. Daniel, and L. M. Silveira. Guaranteed passive balancing transformations for model order reduction. *IEEE Trans. Computer-Aided Design of Integrated Circuits and Systems*, 22:1027–1041, 2003.

[50] R. Pinnau. Model reduction via proper orthogonal decomposition. In W. H. A. Schilders, H. A. van der Vorst, and J. Rommes, editors, *Model order reduction: theory, applications, and research aspects*, pages 95–109. Springer, 2008.

[51] J. W. Polderman and J. C. Willems. *Introduction to mathematical systems theory: a behavioral approach, Texts in Applied Mathematics 26.* Springer, 1998.

[52] P. Rabiei and M. Pedram. Model order reduction of large circuits using balanced truncation. In *Proceedings of the ASP-DAC Design Automation Conference*, volume 1, pages 237–240, 1999.

[53] P. Rabier and W. Rheiboldt. A general existence and uniqueness theory for implicit differential algebraic equations. *Diff. Int. Equations*, 4:563–582, 1991.

[54] M. J. Rewieński. *A trajectory piecewise-linear approach to model order reduction of nonlinear dynamical systems.* PhD thesis, Massachusetts Institute of Technology, 2003.

[55] M. J. Rewieński and J. White. A trajectory piecewise-linear approach to model order reduction and fast simulation of nonlinear circuits and micromachined devices. *IEEE Trans. CAD Int. Circ. Syst.*, 22(2):155–170, February 2003.

[56] W. J. Rugh. *Linear system theory.* Prentice-Hall, 1996.

[57] M. Saadvandi. Passivity preserving model reduction and selection of spectral zeros. Master's thesis, Royal Institute of Technology KTH, 2008.

[58] W. H. A. Schilders. Introduction to model order reduction. In W. H. A. Schilders, H. A. van der Vorst, and J. Rommes, editors, *Model Order Reduction: Theory, Research Aspects and Applications*, pages 3–33. Springer Berlin Heidelberg, 2008.

[59] D. E. Schwarz and C. Tischendorf. Structural analysis of electric circuits and consequences for MNA. *International Journal of Circuit Theory and Applications*, 28:131–162, 2000.

[60] E. D. Sontag. *Mathematical control theory.* Springer, Berlin, 1990.

[61] D. Sorensen. Passivity preserving model reduction via interpolation of spectral zeros. *System and Control Letters*, 54(4):347–360, 2005.

[62] J. Stoer and R. Bulirsch. *Introduction to Numerical Analysis.* Springer, New York, 2nd edition, 1993.

[63] M. Striebel. *Hierarchical Mixed Multirating for Distributed Integration of DAE Network Equations in Chip Design.* Number 404 in Fortschritt-Berichte VDI Reihe 20. VDI, 2006.

[64] M. Striebel and J. Rommesl. Model order reduction of nonlinear systems in circuit simulation: status and applications. In P. Benner, M. Hinze, and J. ter Maten, editors, *Model Reduction in Circuit Simulation*, Lecture Notes in Electrical Engineering. Springer, 2010. Proceedings of SyreNe workshop in Hamburg 2008.

[65] T. Stykel. *Model reduction of descriptor systems*. Institut für Mathematik TU Berlin, 2001. Technical Report 720-2001.

[66] T. Stykel. Balancing-related model reduction of circuit equations using topological structure. In P. Benner, M. Hinze, and J. ter Maten, editors, *Model Reduction in Circuit Simulation*, Lecture Notes in Electrical Engineering. Springer, 2010. Proceedings of SyreNe workshop in Hamburg 2008.

[67] C. Tischendorf. *Solution of index-2-DAEs and its application in circuit simulation*. PhD thesis, Humbold-Univ. zu Berlin, 1996.

[68] C. Tischendorf. Topological index calculation of DAEs in circuit simulation. *Surv. on Math. Ind.*, 9(8):187–199, 1999.

[69] D. Vasilyev, M. Rewieński, and J. White. A TBR-based Trajectory Piecewise-Linear Algorithm for Generating Accurate Low-order Models for Nonlinear Analog Circuits and MEMS. In *Design Automation Conference*, pages 490–495, June 2003.

[70] A. Verhoeven. *Redundancy reduction of IC models by multirate time-integration and model order reduction*. PhD thesis, Technische Universiteit Eindhoven, 2008.

[71] A. Verhoeven, J. ter Maten, M. Striebel, and R. Mattheij. Model order reduction for nonlinear IC models. In A. Korytowski, K. Malanowski, W. Mitkowski, and M. Szymkat, editors, *System Modeling and Optimization, IFIP AICT 312*, volume 312 of *IFIP Advances in Information and Comunication Technology*, pages 476–491. Springer, 2009.

[72] E. I. Verriest. Time variant balancing and nonlinear balanced realizations. In W. H. A. Schilders, H. A. van der Vorst, and J. Rommes, editors, *Model Order Reduction: Theory, Research Aspects and Applications*, pages 231–250. Springer Berlin Heidelberg, 2008.

[73] T. Voß. Model reduction for nonlinear differential algebraic equations. Master's thesis, University of Wuppertal, 2005.

[74] K. Willcox, J. Peraire, and J. White. An Arnoldi approach for generation of reduced-order models for turbomachinery. *Computers and Fluids*, 31(3):369–389, 2002.

[75] Y. Yue and K. Meerbergen. Accelerating optimization with model order reduction. in preparation, 2010.

[76] K. Zhou, J. Doyle, and K. Glover. *Robust and Optimal Control.* Prentice Hall, 1996.